SEEING WAR THROUGH
GOD'S EYES

SEEING WAR THROUGH
GOD'S EYES

JERRY A. WEBSTER

Seeing War Through God's Eyes
Copyright © 2016 by Jerry Webster. All rights reserved.

No part of this publication may be reproduced, stored in a retrieval system or transmitted in any way by any means, electronic, mechanical, photocopy, recording or otherwise without the prior permission of the author except as provided by USA copyright law.

Scripture quotations are taken from the Holy Bible, King James Version, Cambridge, 1769. Used by permission. All rights reserved.

This book is designed to provide accurate and authoritative information with regard to the subject matter covered. This information is given with the understanding that neither the author is engaged in rendering legal, professional advice.
Since the details of your situation are fact dependent, you should additionally seek the services of a competent professional.

Cover design by Jerry Webster
Interior design by Shieldon Alcasid

ISBN: 9780692080344
1. Religion / General
2. Religion / Reference
16.06.02

For my God, my wife, my family!

He that dwelleth in the secret place of the most High shall abide under the shadow of the Almighty. I will say of the Lord, He is my refuge and my fortress: my God; in him will I trust. Surely he shall deliver thee from the snare of the fowler, and from the noisome pestilence. He shall cover thee with his feathers, and under his wings shalt thou trust: his truth shall be thy shield and buckler.

—Psalm 91:1–4

Contents

Introduction 11

1   What Causes War? 15
2   Is It a Just War? 45
3   How Does God Feel about War? 69
4   Why Does God Allow War? 85
5   Gearing Up 103
6   Fighting the Good Fight 119
7   Identifying the Enemy 133
8   Life as a Prisoner of War 153
9   The Final Battle 165
10   Destination Paradise 179

Notes 189

Introduction

FOR YEARS, I was haunted by everything that I had taken part in while I served in the United States Army during combat deployments. The Army trains soldiers actively to go into the combat field and fight using skills in everything that we do to eliminate enemy combatants. Unfortunately, they do not train anyone on getting past all the demons of taking the life of the enemy or getting past destroying innocent lives or taking the life of a young child because of being close to those military targets. Insurgents also use these children to attempt to throw the military off by using them to destroy convoys and patrols to win for their cause.

Growing up in small-town USA, I was raised in a church family and taught "Thou shalt not kill!" to the fullest. I joined the Army in 1986 and started my quest in fighting for my great country. At the time, I never thought twice about what I was doing; however, once I was done, it was hard to live with myself, knowing what took place in the battlefield. When we go to war, we must come to terms with the fact that someone is most likely going to die, and I choose for it not to be me. God can give you assurance to ease your mind from fighting in the battlefields.

In my first book *What Brings a Soldier to His Knees*, I talk about overcoming the demons of combat. In *Seeing War though God's Eyes*, I go deep into the Scripture and show how God, in the Old Testament, supported and ordered combat at times. When one starts to relive their time over and over again, it seems to start to get deep down in the mind. Most of the time, unfortunately, some feel there is only one way out.

I have learned through time that we as combat soldiers, I feel from my own experience going to war, I did what at times seemed to be very wrong. Going on patrols and combat maneuvers, you kill anyone who is in the way that is enemy personnel. We then get home expecting from those we are close to—family, friends, etc.—to condemn us for the actions that we took while downrange. We only find that coming home, we are praised and thanked for a job well done. After no one condemns us as we feel we deserve, we feel the only way to find that peace is to commit suicide, thinking it will all go away. Only problem is, there was nothing wrong with what we did, and after committing suicide, we only transfer the pain to someone else who loves us very much.

We are seeing twenty-two veterans per day taking their own lives. This is twenty-three too many. The time to stop the statistics that we see every day starts now. I say *twenty-three* because twenty-two are committing suicide, and there is always one who is still thinking of doing it. God is our only hope of calming our minds and finding true peace. My prayers are that through my two books, veterans will find aid and find their way back to a happy life, staying Army strong!

1

What Causes War?

WHEN WE TALK about war, the key question is, What causes it? You can find many references in the Bible linked to war and battle. The Bible is the most useful book in the world and can answer any question we ask. Just keep in mind that one needs to be willing to accept the response they get to the main question that we are asking: what causes war or the organization of armed and often prolonged conflict carried on between states, nations, or other parties typified by high violence, social disorder, and increased humanity. First, we must understand that war is defining an existing intentional and widespread protective conflict between government communities and, therefore, is a form of political violence.

I cannot stress enough the importance of understanding the reason that lies behind conflict; note, if you have lack the ability to enforce or commit to an agreement, then a war may last a long time or until one side becomes victorious and successful, the status quo changes, or the costs of continued conflict become too high on all sides. It is possible to have statements both enforceable and credible; states start with unbalanced facts, for instance, about the comparative strength of one of two countries.

In such cases, a bargaining failure can lead to war. In such settings, once war begins, the relative strengths of countries can become clearer.[1] Given that credible bargaining is possible and can avoid further costs of war, the states could then reach an agreement to end the war. Different durations of wars can match to unusual sources of bargaining failures.

Revenge is another reason for war that they would instinctively place within the set of irrational explanations of war.[2] Distinguishing a sentimental version of what someone calls revenge of a repeated game, the punishment phase involved in trigger strategies of one kind or another. The impassioned version that falls within the non-rationalist explanations. Motivated by anger for past actions and not clear drive, considering penalties are to be revenge in emotional terms, nor decided from before as part of a best approach. Wars driven by revenge are real but rare. famous examples include the Achaeans in the Trojan War, at least according to the description in the Iliad. Let's look deeper than the surface in what causes war, international force, or intercontinental conflict. What aggravates and triggers the war? The following are some of the causes of war:

- Opposing interests and skills
- Contact and salience
- Significant change in the balance of powers
- Individual opinions and expectations
- A disrupted structure of expectations
- A will to conflict

The 9/11 attack was a complex international maneuver, the result of years of planning. Bombings like Bali in 2003 or Madrid in 2004, while able to take hundreds of lives, could have been planned locally. Their needs were far more modest in size and complexity.

Terrorists are difficult to prevent than soldiers training defensively. The United States government must set up the safeguards to prevent a 9/11-scale plot from succeeding, and those safeguards must increase significantly to cope with minor but still devastating attacks in the future. A complex international terrorist attack aimed at launching a disastrous attack cannot be mounted by just anyone in any place. For individuals to carry out successful attacks the following must be met:

- Time, space, and ability to perform competent planning and staff work
- A command structure able to make necessary decisions and take over the authority and contacts to assemble needed people, money, and materials
- Opportunity and space to recruit, train, and select operatives with the needed skills and dedication, providing the time and structure required to socialize them into the terrorist cause, judge their trustworthiness, and hone their skills
- A securable logistics network able to manage the travel of operatives, move money, and transport resources (like explosives) where they need to go
- Access, with certain weapons, to the materials needed for a nuclear, chemical, radiological, or biological attack
- Reliable communications between coordinators and operatives and an opportunity to test the workability of the plan

If you look, you'll see several causes of war. When you dig deeper into conflicts one finds what fuels the causes of war and intensifying it through these following issues:

- Sociocultural dissimilarity
- Cognitive imbalance

- Status difference
- Coercive state power

> War arises because of the changing relations of several technological, psychic, social, and intellects. There is no single cause of war. Peace is the equilibrium between forces. Change in any force, trend, movement, or policy may at one time make for war, but under other conditions, a similar change may make for peace. At distinctive times states can promote peace by armament or time by disarmament. At unusual times by the insistence on its right and another time by spirit conciliation. Estimating the likelihood of war takes an appraisal of the effect of current changes on the complex of inter group relationships throughout the world. (Wright 1965, 1284)

A necessity for understanding the causes and conditions of international conflict behavior, violence, and war is an appreciation that they operate as part of an international social field. They are field forces, conditions, and states. This means that these causes and conditions are interrelated, part of a whole, a process, and an equilibrium. In other words, they operate contextually within the conflict helix.

A structure of expectations is based on a particular balance of powers between states. The balance may shift in time, however, and, aggravated by sociocultural dissimilarity and cognitive imbalance, will produce incongruent expectations. Without such incongruences between two states, there would be no conflict situation. There would be no mutual antiforeign riots or demonstrations and tension, friction, and coolness in relations.

In Operation Iraqi Freedom, most people would agree the roots of Iraqi Freedom go back to the Gulf War, regardless of what we think the government's case was. Someone may say the real argument is oil, pride, weapons of mass destruction, or the failure to uphold dozens of United Nations resolutions. They go back to Iraq's invasion of Kuwait, a sovereign nation with treaties with the United States and other coalition forces. It only took fifteen years, from 1950–1965 for the United States to get incrementally involved in the conflict in Vietnam. We started off aiding the French, and before we knew it, we had advisors, then Special Forces, then the 1964 Tonkin Gulf resolution. What do we call that? War by accident?

James D. Fearon (1995) argues that conflict or war typically happens because of anarchy. An international-level example can be where there is no single leader or central authority who would suppress the political or military ambitions of individual states. Involving two or more states or countries describes the gist of conflicts. The arms race during the Cold War is a good example. This competition can easily result in war since each nation state tries to preserve its security. This can result in an upward spiral of ever-increasing military expenses since each state feels the need to match the other's power to ensure its survival. This implies that one state's security always results in another state's insecurity.

An important theory has an ambivalent position about war. On one hand, the crucial theory justifies war, as long as fighting is by suppressed people against their suppressors. Most representatives of the critical theory see conflict as something positive because it helps to break up old political organizations. Examples of such "liberation wars" are the independence wars of former colonies or the "revolutionary war" of the proletariat against the capitalist. On the other hand, the critical theory argues that violent dominance of elites or ruling classes is fundamentally bad, criticizing their fight for dominance and describing it as exploitation.

Contrasting realism and liberalism, critical theory argues actions of war explained by conflict between elite and the rest. These conflicts can be over political power, economic well-being, and resource sharing. An example where a country has tried to economically benefit from going to war is the first Opium War (1839–1842), in which England attacked China to open a market for its profitable opium trade.

God does not cause war, but evil in the land does. Conflicts that we have weathered and evil hitting land is the author of war and have started because of sin.

> From whence come wars and fighting's among you? Come they not hence, even of your lusts that war in your members? Ye lust, and have not: ye kill, and desire to have, and cannot obtain ye fight and war, yet ye have not, because ye ask not. (James 4:1–2)

Look throughout time and the conflicts the United States has been in and examine them. Look for a moment at World War I. What went wrong there and caused the war to take place? We had the Triple Alliance: Germany, Italy, and Austria-Hungary versus The Triple Entente: Britain, France, and Russia. The United States joined on the side of the Triple Entente. The main cause of this war was assassinating Archduke Franz Ferdinand. With this event taking place, it pulled several nations together to war against the other ones. Peace was not taking place, but nations stood for what they believed in and felt was right.

A sufficient cause of conflict is one whose occurrence produces conflict. There is only one such cause, and it is of a conflict situation, not formal or official conflict behavior. This is a significant change in the balance of powers that is in the interests, capabilities, and/or wills of one or both parties.

Such change therefore has a dual effect. It produces a conflict situation, perhaps manifested in tension, hostility, friction, coolness, and antiforeign demonstrations. Interstate relations remain "correct," but beneath, the pot is boiling.

And this change is a necessary cause for the subsequent conflict behavior once expectations have been disrupted.

While affecting some nonviolence also, most of the remaining aggravators primarily act on violence. First of these is big power intervention in the conflict, which may transform a local dispute into one involving the status quo among the powers and thus raise the stakes at issue. Such intervention also injects into the conflict greater resources for confrontation.

More than seventy million military personnel, including sixty million Europeans, were mobilizing in one of the largest wars in history, resulting in killing more than nine million combatants, largely because of extraordinary technological advances in firepower without matching advances in mobility. It was the sixth deadliest conflict in world history.[3]

World War II we had nationalistic tensions, unresolved issues, and resentments resulting from World War I and the interwar period in Europe. Effects of the Great Depression, they thought, were a contributing cause, as well. World War II had the Axis Powers: Germany, Italy, and Japan versus Major Allied Powers: United States, Great Britain, France, and Russia, resulting in the deaths of over sixty million people, making it the deadliest conflict in human history. This war, had several reasons, but the biggest part or cause was again individual interest that had clashed.

Most American wars have obvious starting points or precipitating causes. The Battles of Lexington and Concord in 1775, the capture of Fort Sumter in 1861, the attack on Pearl Harbor in 1941, and the North Korean invasion of South Korea in June 1950 are examples. However, there was no fixed beginning for the United States war in Vietnam. The United States entered that war incrementally, in a series of steps between 1950 and 1965.[4] In May 1950, President Harry S. Truman sanctioned a modest program of economic and military aid to the French, who was fighting to keep control of their Indochina colony, including Laos and Cambodia, as well as Vietnam. When the Vietnamese Nationalist (and Communist-led) Viet Minh army defeated French forces at Dien Bien Phu in 1954, it compelled the French to comply with creating a Communist Vietnam north. Of the seventeenth parallel while leaving a noncommunist entity south of that line, the United States refused to accept the arrangement.[5]

The presidential administration accepted instead to build a nation from the false political entity that was South Vietnam by creating a government there. Taking over control from the French, sending military advisers to train a South Vietnamese army, and unleashing the Central Intelligence Agency to conduct psychological warfare against the North.

With the larger fundamental and political causes of the war in Vietnam, the experience, character, and personality of each president played a role in deepening the United States obligation. Dwight Eisenhower restrained United

States involvement because having commanded troops in battle, he doubted the United States could fight a land war in Southeast Asia. The youthful John Kennedy, on the other hand, felt he had to prove his resolve to the American people and his Communist adversaries, especially in the aftermath of several foreign policies' blunders early in his administration. Lyndon Johnson saw the Vietnam War as a test of his mettle, as a Southerner and as a man. He encouraged his soldiers to "nail the coonskin to the wall" in Vietnam, likening victory to a successful hunting expedition.[6]

Today, we have our war on terrorism that we have fought since September 11, 2001; this is the day that all Americans will remember. When the World Trade Center towers where hit and hitting the Pentagon happened, it was because one group did not like what America was doing. Really, it was not a group, but one man who completely controlled that group. His hatred for the United States was because of the presence of the United States in the Middle East when we fought in the Gulf War.

Seeing War Through God's Eyes   27

We could consider him as a Muslim extremist, fanatical religiosity, hard-line interpretation of Islam. After 9/11, the leaders, seemingly sharing everyone's conviction in the need of decisive retaliation, promised to have done everything that they could do to make America safe in terrorist attack and ensure that it would never happen again. In a universally applauded speech, President Bush pledged to wipe out the enemy by waging a war that was to begin with Al-Qaeda and the Taliban. In the same speech, President Bush vowed to the fight would "not end until every terrorist group of global reach had been found, stopped and defeated."

President Bush vowed,[7] "I will not yield, and I will not rest; I will not relent in waging this struggle for freedom and security for the American people."[8] Fulfilling the promise to defeat the terrorist enemy that struck on 9/11, United States leaders would first need to identify who exactly the enemy was and then be willing to do whatever is necessary to defeat him. Let us examine what this would entail and compare it with the actions the leaders took. Who is the enemy that attacked America September 11? It is not "terrorism," just as the enemy, in World War II, was not a kamikaze strike or U-boat attacks. Terrorism is a tactic employed by a specific group for a definite cause. The group and cause they fight for is America's enemy.

Through time, we can look through each conflict and see the evil that caused that battle. Since beginning of time, we have had conflicts. Our first conflict was Cain and his brother Able in the Old Testament part of the Bible.

> And Cain talked with Abel his brother: and it came to pass, when they were in the field, that Cain rose up against Abel his brother, and slew him. (Gen. 4:8)

This conflict was over a minor issue. Cain was so mad at Able that he slew his own brother.

Just think, the conflict was with family and could easily resolve itself by using the communication method. Some are saying talking does not always work, and they would be right; however, have we quenched all possible ways? I know when we are on the battlefield, one cannot stop combat and talk; nevertheless, anything is likely. Think of the Christmas Truce of 1914 and the story it tells us.

Although the popular memory of World War I is normally one of the horrific casualties and wasted life, the conflict does have tales of comradeship and peace. One of the most remarkable, and heavily mythologized, events concerns the Christmas Truce of 1914. The story goes that soldiers of the western front laid down their arms on Christmas Day and met in no-man's-land. Exchanges were made with food and cigarettes, as well as by playing football. Ending violence was unofficial, and there had been no previous conversation: troops acted spontaneously from goodwill, not orders. Not only did this truce happen, the event was more widespread than commonly portrayed.

You hear many accounts of the Christmas Truce, the most famous of which concerns the meeting of British and German forces; however, French and Belgium troops also took part. The unofficial nature of the truce meant there was no one single cause or origin. Some accounts tell of British troops hearing their German counterparts singing Christmas carols and joining in, while Frank Richards, a private in the Royal Welch Fusiliers, told of how both sides erected signs wishing the other a merry Christmas. From these small starts, some men crossed the lines with their hands up, and troops from the opposing side went to meet them. When officers realized what was happening, the first meetings had been made, and most commanders either turned a blind eye or happily joined in.

The fraternization lasted, in many areas, for all Christmas Day. There was exchanging of food and supplies on a one-to-one basis while in some areas, men borrowed tools and equipment from the enemy to quickly improve their own living conditions. There was playing of games of football using whatever could be found for a ball while burying the bodies that had become trapped within no-man's-land.

Modern retellings of the truce finish with the soldiers returning to their trenches then fighting again the next day, but in areas, the peace lasted much longer. Frank Richard's account explained how both sides avoided from shooting at one another the next day until relieving the British troops, and they left the front line. In other areas, the goodwill lasted for several weeks, bringing a halt to opportunistic sniping before the bloody conflict resumed.

As we talk about what causes war, we learn that several effects can start it, well as fuel it. It is just like a fire; the more wood you use, the bigger the fire. If we look at the meaning of the word, it means extreme aggression, social disruption, and usually high mortality. For example, take the Gulf War; Saddam Hussein while in power was killing the people by the hundreds of thousands. They then invaded Kuwait to take over the land because Iraq wanted revenues from their oil. Greed was causing this war, and where does greed originate? Look and study through time you find the center block of greed is sin.

War emerges from conflict; we can say it is the spark that gets the fire burning. Majority of the time in fighting battles, greed was the individual wanting more. Sad that leaders of this type would send innocent people to do their work and put lives on the line. No sweat to them, they put someone else up front fighting the battle for them so they can get more than they already have. Look back at Cain and Able again; Cain slew his brother because God's blessing was on the offering Able brought and not his. Cain wanted more than his brother.

Looking into your past, at growing up, attending school, and seeing the kids who were always being bullied, and whose stuff was taken from them. We constantly see one consistently wanting everything everyone has. Before you know it, a fight is raging. War starts as a disagreement and then turns into a conflict. The result is war.

It happens everywhere we see conflict on the roads and highways in today's time. Road rage is getting to be at an all-time high. For what? They pulled into my parking spot or cut in front of me while I was driving. We need to go back to the days when God was unfailingly in front of or actions, when he was constantly there, first in our mind. We ask what conflict is. Just look, you have it on your doorstep of your home.

It might be with your neighbor next door or that person next to us at work, but we are in a constant battle ourselves. We fight this world and the evils every day and do not realize how wide it can go. We sometimes think a conflict is just shooting and blowing targets up, but in reality, the true definition is "a fight, battle, or struggle, especially a prolonged struggle; strife. Incompatibility or intrusion, from one idea, want, event, or action with another."

Barbara Ehrenreich tries to answer in her brilliant *Blood Rites: Origins and History of the Passion of War*. Putting it more precisely, what predisposes the human species to get caught up in war fever? Her answer, simply, comes from the fact we are the only species to have gone from mainly prey to almost only predator.

> Here is what we might call the missing link within the theory of human evolution itself: a poor, shivering creature grew to unquestioned dominance. Before and well into the age of hunting, there must have been a long, dark era of fear when careless and stragglers getting picked off routinely, when disease or temporary weakness could turn man into meat.[9]

When Napoleon Bonaparte of France was standing in the battlefield between France and Austria, he was shocked and hurt because the casualties—about fifteen thousand soldiers who died from both sides. He wrote a critical letter to the king of Austria then, saying, "Please let us make peace and let us stop the war." He wrote to the king: "You have not been in the battle like I have, so you do not know what it is to stand here with fifteen thousand corpses around you and the daily suffering of the soldiers, and that is not all. How about those who stay behind?" Yes, I think most of us have watched the film *Born on the Fourth of July*, and we know what it is like. You look at aftereffects also, not only the immediate effects.

So conflicts start brewing right at home before it goes outside. Not handling conflict in the proper ways can intensify into full-scale war. Getting to war, certain issues take place—first, disagreement; after that, conflict; and, subsequently, an all-out war. The greatest knowledge, however, is if we take conflict before God, He will help us to a more peaceful resolution. The problem found man not getting off his high horse and getting on his knees and praying before a God that cares. We need to forget about our pride, then we will see a significant change in the world.

At the present time, our cause for global war on terror is for one reason—the heartbreaking terrorist attacks against the World Trade Center, the Pentagon, and destroying four airliners resulted in around 3,000 deaths. My heart sank, as did millions of people around the world, as rescue efforts tried to find and save lives among the rubble of the destruction. No other external attacks on the United States resulted in so many deaths. By comparison, the Revolutionary War resulted in roughly 4,400 battle deaths, and the attack on Pearl Harbor in 1941 caused nearly 2,300 deaths. If the estimated death toll holds, this makes the September 11 terrorist attacks the greatest death toll on United States soil because of human design since the Civil War.

United States officials claim that at least nineteen Islamic men orchestrated the suicidal hijacking of four Boeing jets that had connections to the fanatical religious leader Osama bin Laden. At least one hijacker owned a copy of the Koran, which authorities found in his bag that did not make the flight. A page had spiritual instructions that contained statements like, "God loves the job you are doing." It, in addition, stated, "When one ends their day in heaven, where you will join the virgins, and I pray to you God, to forgive me from my sins, so I can glorify you in every possible way." The Koran, their faith, and their will to die for their beliefs provide valuable clues for reasons of terrorist actions. Clearly, this clue has yet to affect the minds of military and political leaders.

Holy War or Jihad comes directly from of the heart of fundamentalist Muslims, who base their beliefs on the Koran. Receiving heavenly rewards originated from a legend of a Shiite leader, Hassan-I Sabbah, of the sect at Alamut (Alamut served as a fortress in Persia less than ten years before the First crusade). He had a private garden of infinite beauty, with the sparkling fountain so precious to the desert dweller, and a selection of the most beautiful and most sexually succeeded young women in the land.

A youthful member of the sect received hash to numb his mind to the point of unconsciousness. When he awoke, he found himself in the fabled garden, where the junior women fed him morsels of the most delicious foods. They treated him to every sexual delight he had ever heard of and to some that he had never imagined.

When he awoke the next morning to his familiar surroundings, he recounted his experience. Someone told him that he had received a gift from Allah by allowing him a glimpse, of the highest level of heaven reserved for those martyrs who die for their faith. The delights he had experienced so briefly would happen to him through all eternity if he met the standard for his faith. He begged for nothing more in the world but dying in service for Allah. In response to his plea, he received intense training to defeat an enemy of God, identified by his leader, a grand master.

His training included the techniques of the dagger, where and how to strike and how to evade armor (sound familiar?). Thus, he would receive everlasting bliss in paradise for his suicide mission. Based on their alleged use of hashish to trick young followers, the sect became known as the Hashshashin. This proved a complex world for the Crusaders to pronounce so they took the closest Church Latin term, *Assassini*, which later changed into the English form, *assassins*.

Today, extremist Islamic terrorist organizations have formed cells in many countries throughout the world, and they come from both Sunni and Shiite factions. All these groups usually share a fundamentalist Islamic belief that adopts holy war against persecutors and nonbelievers who fight against Islam. The Islamic Jihad recruits teenage men by religious indoctrination and bases its terror tactics on the willingness of these young people to lay down their lives for what they see as a divine command. They honestly believe that their god will reward them in heaven. They perform suicides, destroying people's lives. They carry these duties wholeheartedly not from dictatorship but from free will to follow and trust in their God.

The belief in the Koran with Jihad serves as just one example for the defense of terrorist attacks. Their religious beliefs point to the cause of terrorism. Unfortunately, the political leaders either do not understand this or refuse to recognize this real clear point because to accept a belief as a cause hits at their own religion and theological justification. The violent injunctions of the Quran and the violent precedents set by Muhammad set the tone for the Islamic view of politics and of world history. Islamic scholarship divides the world into two spheres of influence, the House of Islam (dar al-Islam) and the House of War (dar al-harb). Islam means submission, and so the House of Islam includes those nations that have submitted to Islamic rule, which is to say those nations ruled by Sharia law. The rest of the world, which has not accepted Sharia law and so is not in a state of submission, exists in a state of rebellion or war with the will of Allah. It is incumbent on dar al-Islam to make war upon dar al-harb until such time that all nations submit to the will of Allah and accept Sharia law. Islam's message to the non-Muslim world is the same now as it was in the time of Muhammad and throughout history: submit or be conquered. The only times since Muhammad when dar al-Islam was not actively at war with dar al-harb were when the Muslim world was too weak or divided to make war effectively.

Let's look at a couple sample verses from one of the more benign translated versions of the Koran:

> And kill them wherever you find them, and drive them out from whence they drove you out, and persecution is severer than slaughter, and do not fight with them at the Sacred Mosque until they fight with you in it, but if they do fight you, then slay them; such is the recompense of the unbelievers. (Koran 2:191)

> Fight them, Allah will punish them by your hands and bring them to disgrace and assist you against them and heal the hearts of a believing people. (Koran 9.14)

> Or do they desire a war? But those who disbelieve shall be the overthrown ones in war. (Koran 52.42)

Whatever serves as a description for their acts, *cowardly* simply does not fit. *Criminal*, *inhumane*, *disgraceful*, *practical boldness*, or *fanatical* give far better portrayals, but surely not *cowardly*. Nor did these fanatical terrorists take on to remain faceless. All of them left obvious and intended traces to their identities. After merely a few days, investigators claimed to have discovered their names and their photos. Most people will agree these terrorists acted out of fanaticism, but remember that *fanatic* serves only solely as a relative term for a person who holds more zeal in their beliefs than us.

Many claimed that one purpose of history aims to prevent the errors of past events, yet we had to continue the same rhetoric used by war starters throughout history. President George W. Bush has resorted to ancient Bronze Age thinking and appealed to biblical beliefs of war, punishment, and evil. President Bush vowed that the United States would find and punish "those behind these evil acts" and any country that harbors them. Even the Defense Secretary Donald Rumsfeld stated his desire to "punish the perpetrators, as soon as they are identified."

Note that within the United States, we claim to live by the rule of law. In the United States, we treat criminals and terrorists by the justice of law, but outside of the United States, we treat with punishment or war. *War* serves as a word that justifies murderous actions outside the rule of law, with a religious belief to serve as its main justification.

President George Bush declared September 14 as a day for prayer and remembrance to prepare the minds of US citizens for the future. President Bush asked the nation to pray for the families of the victims and quoted the book of Psalms, "And I pray they will be comforted by a power greater than any of us spoken through the ages in Psalm. 'Even though I walk through the valley of the shadow of death, I fear no evil for you are with me.'" The implied meaning, of course, puts God on the side of the United States. Remember, the Islamic terrorists feel that God stands on their side.

President George Bush stated, "This crusade, this war on terrorism is going to take a while," as he spoke to the reporters as a medieval warlord. This dangerous and not-so-thoughtful comment stoked suspicion in Arab and Muslim quarters, where *crusade* is a loaded term that recalls the Christian wars against Muslims in the Holy Land. President Bush had to apologize for the statement, but apologies come cheap, and many Islamic people will not have heard his apology, nor believe him, even if they heard it.

Beliefs connected to words have consequences. Who knows how many fence-sitting Muslims will side with terrorists simply due to a careless word, a word that has weight in the lives of civilians and military men? On September 21, Sheikh Maher Hammoud responded with, "It is a war against Islam even though the White House had apologized for President Bush's statements of them being a crusade."

On September 24, Osama bin Laden, addressing the defenders of Islam, stated, "We hope that these brothers are among the first martyrs in Islam's battle, against the new Christian-Jewish crusade led by the big crusader President Bush under the flag of the Cross, this battle will be considered one of Islam's battles."

What the military does not like to specify, of course, is the horrific fact that the secondary damage includes not only civilian buildings but the innocent human beings who live in or around them. When asked about the extent of Iraqi casualties at the end of the Gulf War, then-military chief of staff Colin Powell blandly remarked, "That is not a matter; I am terribly interested in."[10]

Interestingly, Timothy McVeigh, the Oklahoma City bomber who the United States executed (but not punished) though the justice system, had incorporated the term from his Gulf War experience as a soldier. In an interview before his execution, he outraged the American public when he described the 19 dead children among his 168 victims as "collateral damage." How easy to feel outraged when collateral damage goes against Americans, but, sadly, the same does not apply when the government does the same destruction to people in other countries.

The Saudi-born millionaire, Osama bin Laden, served as a veteran of the 1979–1989 Afghan war against the Soviets, where he came to "understand" the conflict because of

Muslim versus heretics based on the Koran. Interestingly, in the late 1980s, the United States government supplied $500 million worth of arms to Muslim fighters in Afghanistan, including Osama bin Laden and the Taliban rulers. The arms included high-powered sniper rifles and shoulder-fired stinger antiaircraft missiles.

For him to further his religious "duties," bin Laden founded the "International Islamic Front for Jihad against the Jews and the Crusaders." Bin Laden's name has come up in several terror attacks around the world, among them the attacks in Riyadh (November '95) and Dhahran (June '96) that left about 30 people dead, including 24 Americans.

Some also associated him in the attacks on a Yemenite hotel (December '92) that injured several tourists; the assassination tried on Egyptian President Mubarak in Ethiopia (June '95); the World Trade Center bombing, as well, (February '93) that killed 3 and injured hundreds; the Somali attack on American forces that left hundreds wounded; and the Yemen attack on the USS Cole (October 2000) that killed 17 sailors and injured 39. Note that all these attacks occurred after the Gulf War.

The causes of war are to show your power over another country, but there are many different reasons war starts. First, nations think it helps meet their problems and influential political issues. Never knowing why, people start fighting with another country. Even when they are aware that it will bring destruction to both countries, they still go ahead and develop war, not realizing they are putting people's lives in danger.

Finally, wars can be caused by racial conflicts or the fight for independence. This occurs when wars are fought by suppressed people against their rulers or foreign occupations. In the 1950s and 1960s, many former colonies of the British and French empires fought against colonial powers, such as in the Algerian War of Independence against France.

Sometimes social conflicts within societies can be characterized as a class war or racial war. The war of the black majority against the system of Apartheid in South Africa is an example of racial conflict. Apartheid was an institutionalized system of racial segregation enforced by the National Party (Nasionale Party) government in South Africa from 1948 to 1994. One difference between South Africa's apartheid era and other periods of racial segregation that have occurred in other countries is the systematic way in which the National Party formalized it through law.

The French Revolution is another example, where the middle and lower classes fought against the French aristocracy. The Revolution took shape in France when the controller general of finances, Charles-Alexandre de Calonne, arranged the sending for of an assembly. The sending of "notables" in February 1787 to propose reforms designed to remove the budget shortage by increasing taxing the privileged classes. The assembly refused to take responsibility for the reforms and suggested calling the estates general, which represented the clergy, the nobility, and the Third Estate (the commoners), which had not met since 1614.

Usually, you get a mixture of motives and reasons for every war. To understand the reasons for why wars occur, one needs to distinguish between the individual and collective motives. A psychological idea, such as narcissism, is another possible approach to understanding why individuals fight. This is related to the desire of many men to achieve glory and individual honor by fighting a war. These psychological theories of war argue that men can increase their social status by fighting in a state of war and coming back as war heroes.

2

Is It a Just War?

WHAT CAUSES WAR? This question has been answered in the previous chapter, but the range and nature of all the causes and conditions may not be clear because the discussion moved across phases and subphases of conflict and types of causes.

War is generated by a field of sociocultural forces seated in the meaning, values, and norms of states. Specifically, war is an outcome of an imbalance among these forces in international space-time and is the process through which a new field equilibrium is established.

The causes and conditions of war, therefore, operate within this social field. They are interrelated; their operation is relative to the space-time. War is therefore not the product of one cause or x number of causes operating independently. War is a social field phenomenon, and its causes and conditions must be understood as aspects of this field as contextual, situational.

Authoritative Catholic Church teaching is where we see the just war theory, confirmed by the United States Catholic Bishops in their pastoral letter, The Challenge of Peace: God's Promise and the response, issued in 1983. More recently, the Catechism of the Catholic Church, in paragraph 2309 has listed four strict conditions for "legitimate defense by military force."

1. Damage inflicted by the aggressor on the nation or community of nations must be lasting, grave, and certain:

2. All other means of stopping it must have been shown to be impractical or ineffective.

3. There must be serious prospects of success.

4. The use of arms must not produce evils and disorders graver than the evil to be removed. The power, as well as the precision of modern means of destruction, weighs heavily in evaluating this condition.

Moral hatred toward war with a willingness to accept that war may sometimes be necessary was combined by just war theorists. You see, there had to be legitimate defenses by military force for war to be just. Before going into battle, we need to ensure that we have a genuine defense

Seeing War Through God's Eyes

by the military by going through the list. If we have no appropriate excuse, then it's time to stop elevation of war. The principles of the just war theory must be upheld.

- A just war can only be waged as a last resort. All nonviolent alternatives must be exhausted before the use of force can be justified.

- A war is just only if it is waged by a legitimate authority. Even just causes cannot be served, by actions taken, by individuals or groups who do not make up an authority sanctioned, by whatever the society, and outsiders to the society view legit.

- A just war can only be fought to redress a wrong suffered. For example, self-defense against an armed attack is always considered being a cause justified, although the justice of the cause is not enough to see fourth point. Further, a just war can simply be fought with "right" goals the only permissible objective of a just war is to redress the injury.

- A war can only be just if it is fought with a reasonable chance of success. Deaths and injuries incurred in a hopeless cause are not morally justifiable.

- The final goal of a just war is to reestablish peace. Specifically, the peace set up after the war must be preferable to the peace that would have prevailed if the war had not been fought.

- The violence used in the war must be proportional to the injury suffered. States are restricted from using force not needed to arrive at the limited objective of addressing the injury suffered.

- The weapons used in war must be discriminate between combatants and noncombatants.

- Civilians are never permissible targets of war, and every effort must be taken to avoid killing civilians. The deaths of civilians are justified only if they are unavoidable victims of a deliberate attack on a military target.

It has been many years since September 11, 2001—the day that Islamic terrorists incinerated thousands of innocent individuals in the freest, wealthiest, happiest, and most powerful nation on earth.

On that day and in the weeks after, we all felt the same things. We felt grief that we had lost so many who had been so good. We felt anger at whoever could commit or support such an evil act. We felt disbelief that the world's only superpower could let this happen. And we felt fear from the newfound realization that such evil could rain on any of us. But above all, we felt the desire for overwhelming retaliation against whoever was responsible for these atrocities, directly or indirectly, so that no one would dare launch or support such an attack on America ever again.

To conjure up the emotions we felt on 9/11, many intellectuals claim, is dangerous, because it promotes the "simplistic" desire for revenge and casts aside the "complexity" of the factors that led to the 9/11 attacks. But, in fact, the desire for overwhelming retaliation most Americans felt after 9/11—and feel rarely, if ever, now—was the result of an objective conviction: that a truly monstrous evil had been perpetrated and that if the enemies responsible for the 9/11 attacks were not dealt with decisively, we would suffer the same fate (or worse) again.[1]

The group that threatens us with terrorism—the group of which Al-Qaeda is but one terrorist faction—is a militant, religious, ideological movement best designated as Islamic totalitarianism. The Islamic totalitarian movement, which enjoys widespread and growing support throughout the Arab-Islamic world, encompasses those who believe that all must live in total subjugation to the dogmas of Islam and who conclude that jihad (holy war) must be waged against those who refuse to do so. Islamic totalitarians regard the freedom, prosperity, and pursuit of worldly happiness animating the West (and especially America and Israel) as the height of depravity. They seek to eradicate Western culture, first in the Middle East and then in the West itself, with the ultimate aim of bringing about the worldwide triumph of Islam. This goal is achievable, adherents of the movement believe, because the West is a "paper tiger" that can be brought to its knees by sufficiently devoted Islamic warriors.

Without physical and spiritual support by these states, the Islamic totalitarian cause would be a hopeless, discredited one, with few, if any, willing to kill in its name. Thus, the first order of business in a proper response to 9/11 would have been to end state support of Islamic totalitarianism, including ending the Iranian regime that is its fatherland. As a secondary priority, a proper fight against the enemy that attacked on 9/11 would have involved ending state sponsorship of terrorism by Arab states derivatively connected to Islamic totalitarianism—states such as Syria (and, before it was ended, Saddam Hussein's Iraq). These regimes are active supporters of Arab-Islamic terrorism and mouth support for the Islamic totalitarian cause but are not ideologically committed to it; these regimes support this cause out of political expediency. Supporting Islamic totalitarianism gains power for them; by supporting anti-Western causes and jihadists, Arab states direct the misery of their people toward America and Israel and away from their own brutal rule. Supporting Islamic totalitarianism also gains money for Arab states—for example, the leaders of Syria, a stagnant nation with no oil wealth, are wealthy because oil-rich Iran pays them for providing assistance to terrorist organizations such as Hezbollah. Dealing effectively with these accessories to Islamic totalitarianism would require, first and foremost, getting rid of the primary supporters of the movement. The next step would be, where necessary, making clear to these derivative regimes that any cooperation with that movement or its aims is not expedient but a guarantee of their destruction.

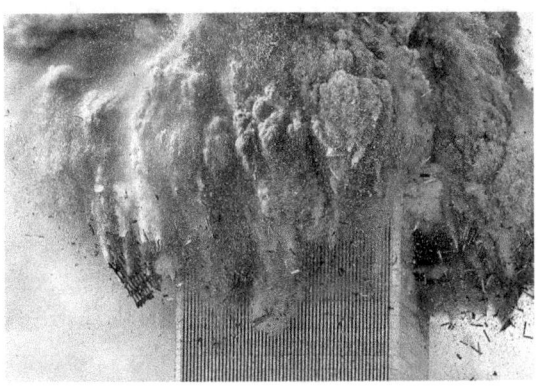

Observe that nearly five years after the terrorist attacks of 9/11—longer than it took to defeat the far more powerful Japanese after Pearl Harbor—the two leading supporters of Islamic totalitarianism and the majority of their accessories remain intact and visibly operative. Iran is aggressively pursuing nuclear weapons, led by a president who declares that our ally Israel must be "wiped off the map," and by Mullahs who lead the nation in weekly chants of "Death to America." Abroad, Iran's terrorist agents kill American troops in Iraq while its propagandists attempt to push Iraq into an Islamic theocracy. Saudi Arabia continues to fund schools and institutions around the world that preach hatred of America and advocate Islamic totalitarianism. Syria remains the headquarters of numerous terrorist organizations and an active supporter of the Iraqi insurgency that is killing American troops. The Palestinian Authority continues a terrorist jihad initiated by Yasser Arafat—a jihad that can be expected only to escalate under the entity's new leadership by the Islamic totalitarian group Hamas. Throughout the Arab-Islamic world, "spiritual leaders" and state-owned presses ceaselessly incite attacks against the West without fear of reprisal.

America has done nothing to end the threat posed by Iran and Saudi Arabia, nor by Syria and the Palestinian Authority. In the rare cases that it has taken any action toward these regimes, its action has been some form of appeasement: extending them invitations to join an "anti-terrorism" coalition (while excluding Israel), responding to the Palestinians' jihad with a promised Palestinian State, declaring "eternal friendship" with Saudi Arabia and inviting its leaders to vacation with our president, and responding to Iran's active pursuit of nuclear weapons with the "threat" of possible eventual inspections by the UN.

Of course, America has done something militarily in response to 9/11; it has taken military action against two regimes: the Taliban in Afghanistan and Saddam Hussein's Iraq. But in addition to these not having been the two most important regimes to target, our military campaigns in each case have drastically departed from the successful wars of the past in their logic, aims, methods, and in their results. In Afghanistan, we gave the Taliban advance notice of military action, refused to bomb many top leaders out of their hideouts for fear of civilian casualties, and allowed many key leaders to escape in the Battle of Tora Bora. And in Iraq, we have done far worse. While we have taken Saddam Hussein out of power, we have not eradicated the remnants of his Ba'athist regime, defeated the insurgency that has arisen, or taken any serious precaution against the rise of a Shiite theocracy that would be a far more effective abettor of Islamic totalitarianism than Saddam Hussein ever was.[2]

I believe two moral conclusions can be made about the present "war." The September 11 attack frames a crime against humanity and cannot be justified. Bombing Afghanistan is also a crime you cannot justify. I have puzzled and questioned the actions our leaders are making and looked over the whole picture. How can war be just when it involves daily killing of civilians and causes thousands of people to leave their homes to avoid the bombs? Let's look into the effects on the innocent when those who planned the September 11 attacks escape the search we set in motion. Ranks of people could increase being angry enough at America for nonstop bombing and combat missions in their country to side with terrorist themselves.

The War on Terrorism amounts to gross violations of human rights and will produce the exact opposite of what we want. It will not end terrorism; it will expand in terrorism. I believe the progressive supporters of the war have confused a just cause with a just war. We have unjust causes—the desire of the United States to set up power in Vietnam, dominate Panama or Grenada, or subvert the government of Nicaragua. A cause may be just getting North Korea to withdraw from South Korea, getting Saddam Hussein to withdraw from Kuwait, or ending terrorism. However, it does not follow that going to war to support that cause, with the unavoidable mayhem that follows, is just.

The city of Kandahar is reported as being a ghost town, more than half of the 500,000 people fleeing the bombs after the attack. The city suffered for seventeen straight days. The city's electrical grid was knocked out, and without electricity, the city was deprived of water because of the pumps could not perform. A sixty-year-old farmer told an Associated Press reporter, "We left in fear of our lives. Every day and every night we hear the roaring and roaring of planes; we see the smoke, the fire…I curse them both, the Taliban and America."

Those were a few weeks following the bombing, and the result had already been to frighten hundreds of thousands of Afghans into abandoning their homes and taking to the dangerous, mine-strewn roads. "War against terrorism" has become a war against innocent men, women, and children, who are in no way responsible for the terrorist attack on New York, yet some people say this is a "just war."

Terrorism and war have a common mission. They both involve killing innocent people to achieve what the killers believe is a complete end. I can see an immediate objection to this equation. They (the terrorists) deliberately kill guiltless people; we (the war makers) aim at "military targets." After killing all the civilians, we start labeling them as "collateral damage" for all the civilians who are by accident.

Can you call it an accident when civilians die under the bombs even if, we grant, the purpose was not to kill civilians? The deaths of civilian people are unavoidable in bombing, not deliberate, but not an accident, and bombers cannot be considered innocent. They are committing murder as surely as the terrorists.

The absurdity of claiming innocence in such cases becomes obvious when the death tolls from "collateral damage" reach figures far greater than the lists of the dead from even the most horrific act of terrorism. Thus, the "collateral damage" in the Gulf War caused more people to die, hundreds of thousands, if we take account of the victims of the sanction's policy, than the mere deliberate terrorist attack of September 11. The number of people who died in Israel from the Palestinian terrorist bombs is somewhere under 1,000. The number of dead from "collateral damage" in bombing Beirut during Israel's invasion of Lebanon in 1982 was roughly 6,000.

We need not match the death lists. It is an ugly exercise thinking one atrocity is worse than another. No killing of innocents, whether deliberate or "accidental," justifies our actions. My argument is with children dying under the hands of terrorists, whether intended or not, because of bombs dropped from airplanes, terrorism and war become equally unpardonable. Let's talk about *military targets*. The phrase is so loose that President Truman, after the atomic bomb wiped out the population of Hiroshima, could say, "The world will note the first atomic bomb was dropped on Hiroshima, a military base. That was because we wished in this first attack to avoid, as possible, killing civilians."

Historically, a just war tradition is a set of agreed rules of combat, may be said to commonly evolve between two culturally similar enemies; that is, when two warring peoples share an array of values, we often find that they implicitly or directly agree on limits to their warfare.[3] However, when enemies differ because of different religious beliefs, race, or language, and they see one another as less than human, war conventions are rarely applied.

When a side sees the enemy as a person who shares similar ethics or traits, they will do business in peace, and there are tacit rules on how a war between them should be fought, as well as who should be involved. Agreeing to these conventions is mutually beneficial and preferable. For instance, both parties understand that the use of underhanded tactics or weapons may provoke an indefinite series of acts of vengeance and thus avoid it.]. Actions like these have proven damaging on political or ethical interests to both sides in the past [, so they avoid it as much as possible.

Regardless of conventions that historically formed, it is the concern of most of the just war theorists the lack of rules to war. They also felt denouncing any asymmetric goodness between belligerents and rules of war should apply to all equally. That is, just war theory needs to be universal, binding on all, capable of appraising the actions of all parties above and beyond any historically formed conventions.

Just war tradition indeed is as old as warfare itself. Early records of joint fighting point out some moral thoughts used by warriors to limit the outbreak or to rein in the probable destruction of warfare. They may have involved consideration of women and children or treating prisoners (enslaving them rather than killing them, or ransoming or exchanging them). Commonly, the earlier behaviors appealed to honor, considering the acts committed in war to be disgraceful.

Honorable is often specific to culture. For instance, a suicide attack or defense is considered honorable by some people but foolish by others. Robinson (2006) notes that honor conventions are also slippery, giving way to practical or military interest when needed.[4] Specifics of honorable differ with time and place. Note that warfare has been infused with some moral concerns from the beginning rather than war being a mere Macbethian bloodbath.

The just war theory also has a long history. Parts of the Bible hint at ethical behavior in war and ideas of just cause. Typically announcing the justice of war by divine intervention, the Greeks may have paid lip service to the gods. As with the Romans, practical and political issues overwhelm any fledgling legal terms—that is, interests of state or Realpolitik (the theory known as political wisdom would take precedence in declaring and waging war.

Nonetheless, this has also been the reading of political realists who enjoy Thucydides' *History of the Peloponnesian War*, an example of why war is to extend politics and from here permeated by hard-nosed state interest rather than "lofty" pretensions to moral behavior.

Since the terrorist attacks on the United States on 9/11 in 2001, academics have turned their consideration to just war once again. With international, national, academic, and military conferences developing and consolidating the theoretical features of the agreements, just war theory has become a popular topic in international relations, political science, philosophy, ethics, and military history courses. Conference proceedings are regularly published, offering readers a breadth of issues the topic stirs—for example, "Human Rights and Military Intervention (Alexander Moseley and Richard Norman, eds), "Just War in a Comparative Perspective" (Paul Robinson, ed.), and War Crimes and Collective Wrongdoing (Aleksander Jokic).

Rules and settlements of warfare remain intact on the battlefield, what has been of great interest is that in the headline wars of the past decade, the dynamic interplay of the rules and conventions of warfare not only remain intact on the battlefield, but their role—and hence their clarification—has been awarded a higher level of scrutiny and debate. In the political circles, justification of war still needs, even in the most critical analysis, a superficial acknowledgment of justification. On the ground, generals have praised their troops to adhere to the rules, teaching soldiers the just war conventions in the military academies (for example, directly through military ethics courses or implicitly through veterans' experiences).

Despite the emphasis on upholding war's conventions, war crimes continue. Genocidal campaigns have been waged by mutually hating people. Leaders have waged total war on ethnic groups within or without their borders, and individual soldiers or guerrilla bands have committed atrocious, murderous, or humiliating acts on their enemy. Just War theory postulates that war, while terrible, is not always the worst option. There may be responsibilities so important, atrocities that can be prevented or outcomes so undesirable they justify war.

The continued brutality of war in the face of conventions and courts of international law lead some to insist that the application of morality to war is a nonstarter, that state interest or military demand would always overwhelm moral concerns. You have people of a more skeptical persuasion who do not believe that morality can or should exist in war; its nature precludes ethical concerns.

Having several ethical viewpoints, there are still a couple of common reasons laid against the need or the possibility of morality in war. People may claim employing all methods in seeking military victory and gaining the victory with a minimum of expense and time. Arguments from military need are of this type. For example, to defeat Germany in World War II, it was necessary to bomb civilian centers, and in the US Civil War, it was necessary for General Sherman to burn Atlanta.

Intrinsic may also decree that no morality can exist in the state of war. They may claim that it can only exist in peaceful situations in which, for instance, recourse exists to conflict resolving institutions. Alternatively, subjectivists may claim that controlling a just cause (the dispute from goodness) is a proper complaint for continuing whatever means are necessary to gain a victory or to punish an enemy.

A different skeptical argument, one advanced by Michael Walzer, says that inventing nuclear weapons alters war so much that our notions of morality and, therefore, just war theories become redundant. However, arguing against Walzer, it can be said that although such weapons change the nature of warfare (for example, the timing, range, and potential devastation), they do not dissolve the need to consider their use within a moral background. A nuclear warhead remains a weapon, and weapons can be morally or immorally employed.

As we examine the just war theory and the principles of the just war, we can examine each war the United States has faced and ask ourselves, Was that a just war or a nonjust war? An acceptable summary of military action might include a preemptive strike too preemptively. Another action to surgically destroy an enemy nation's nuclear weapons production ability with minimum loss of civilian lives, as the Israelis did when they destroyed the Iraqi Osirak nuclear reactor in 1981.

Examples of just wars in US history are World War II, Korea, Vietnam, Desert Storm. Examples of unjust wars and invasions are Somalia, Haiti, Bosnia, Kosovo, and a new unprovoked United States invasion of Iraq. If any war does not properly fall under these categories or wage without provocation as an act of aggression against countries, which has not violated international borders, considering then that war may be "unjust."

In the years past, we can see the United States has had its share of nonjust wars; we need to realize, that we cannot undo these wars and that we should instead live for the present. The main question is, Can we change and learn from past events? We cannot tell leaders how to run the country or missions, but we can seriously talk to God on bended knees, and He will work His part. He has a way to make everything happen when He sees His people call on Him.

Just war theorists combine both a moral abhorrence toward war with a readiness to accept that war may sometimes be necessary. The criteria of the just war tradition act as an aid to deciding whether resorting to arms is morally permissible. Just war theories are attempts to distinguish between justifiable and unjustifiable uses of organized armed forces. They try to understand how the use of arms might be restrained, made more humane, and eventually directed toward the aim of setting up lasting peace and justice.

Appearing early was the notion of the just war, first explained by Augustine, but well-outlined by Aquinas and Grotius. According to the standard theory, if a war is ever to be fought, its principles should be just, keeping in mind both human beings as made in God's image and the reality of human sinfulness:

1. Just cause: The only morally legitimate reason to go to war is for self-defense (or for defending a nation in moral need of defense) or if there is a strong reason for a preemptive strike (for example, a rogue nation with dirty bombs): "If this rule were universally followed there would be no aggressors and no wars."

2. Last resort: "War should be entered upon only when negotiation, arbitration, and compromise, and all other paths fail; for as a rationally being man should, if at all possible, settle his disputes by reason and law, not by force."

3. Lawful declaration: Only a lawful government has the right to begin a war. Simply, the state, not individuals or parties within the state, can legitimately exercise this authority.

4. Immunity of noncombatants: "Those not officially serving as agents of the government in its use of force, including prisoners of war and medical personnel and services, should not be allowed to fight and are not to be subject to violence."

5. Limited objectives: Since the goal of war is peace, not the destruction of the enemy's nation's economy, people or the destruction of its political institutions.

6. Limited means: "Only sufficient force should be used to resist violence and restore peace." *Sufficient* does not necessarily mean decisive victory.

The war on terrorism and voices across the political spectrum, including people on the left, have described this as a just war. One longtime advocate of the peace, Richard Falk, wrote in *The Nation* that this was "the first truly just war since World War II." Robert Kuttner, another consistent supporter of social justice, declared in *The American Prospect* that only people on the extreme left could believe this is not a justifiable war.

Opposing bombing, Afghanistan does not favor giving in to terrorism or appeasement. It asks that we find means other than war to solve the problems that confront us. Martin Luther King and Gandhi both believed in nonviolent direct action, which is more powerful and more morally defensible than war. Pacifism is caricaturing "turn the other cheek." To reject war is not ignoring that war is active; it is, in the present instance, to act in ways that do not mirror terrorists. The United States could have treated the September 11 attack as a horrific criminal act that calls for understanding the culprits, using every device of intelligence and investigation possible. America could have gone to the United Nations seeking to enlist the aid of countries feeling the effects on the pursuit and apprehension of the terrorists.

You have the avenue of negotiations, though the United Stated does not negotiate. The United States did negotiate with bringing and keeping into power monstrous governments throughout the world. Before Bush ordered in the bombers, the Taliban offered to put bin Laden on trial. The president ignored this and started the war against terrorism. After ten days of air attacks, the Taliban called for a halt to the bombing and said they would be willing to talk about handing bin Laden with a third country for trial.

The headline the next day in The New York Times read "President Rejects Offer by Taliban for Negotiations." They quoted President Bush as saying, "When I said no negotiations, I meant no negotiations." Behavior like this is someone steadfast going to war. There were similar rejections of negotiating possibilities at the start of the Korean War, the war in Vietnam, the Gulf War, and the bombing in Yugoslavia. The result was a vast loss of life and incalculable human suffering.

The reality is the term *military* covers all sorts of targets that include civilian populations. When the bombers deliberately destroy, as they did in the war against Iraq, the electrical infrastructure, thus making water purification and sewage-treatment plants inoperable, this will lead to epidemic waterborne diseases; in this case, the deaths of children and other civilians are not accidental. Recall that in the midst of the Gulf War, the United States military bombed an air-raid shelter, killing four hundred to five hundred men, women, and children who were huddling to escape bombs. The claim was that it was a military target, housing a communications center, but reporters going through the ruins immediately afterward said there was no sign of anything like that.

Indeed, in both World War II and Vietnam, the historical record shows there was a deliberate decision to target civilians to destroy the morale of the enemy. From here, there was the firebombing of Dresden, Hamburg, Tokyo, and the B-52s over Hanoi, the jet bombers over peaceful villages in the Vietnam countryside.

When some argue that we can engage in limited military action without an excessive use of force, they are ignoring the history of bombing. The momentum of war rides roughshod over limits. The moral equation in Afghanistan is clear. Civilian casualties are certain. The result is uncertain. No one knew where bombing would lead us, whether it would lead to the end of the Taliban (possibly), a democratic Afghanistan (unlikely), or an end to terrorism (most likely not).

In short, the United States needs to pull back from being a military superpower and become a humanitarian giant, building instead of destroying lives. America needs to be modest, which will lead to a secure and safe future. The modest nations of the world do not face the threat of terrorism. What if a war and all of its suffering could be avoided by highly selective killing? Could just war theory support assassination, for instance? The second principle of just conduct is that any offensive action should remain strictly proportional to the objective needed. Notice that this principle overlaps the proportionality principle of just cause but is distinct enough to be considered in its own light. Proportionality for jus in bello needs tempering the extent and violence of warfare to lessen destruction and casualties.

Assassination programs have been secretly accepted and employed by states throughout the centuries, challenging to a "higher" value such as the self-defense, killing a target guilty of war crimes and atrocities, or removing a threat to peace and stability. The Central Intelligence Agency manual on assassination (1954, cf. Belfield), sought to distinguish between murder and assassination, the latter being justifiable according to the higher purposes sought. This is similar to just war theorists seeking to put mass killing on a higher moral ground than pure massacre and slaughter and fraughting.[5]

Attempting to prevent discrimination, assassination would be conventional if only targets were legitimate, excluding wife or children of a legitimate target. On grounds of the policy would also be acceptable, for if one man or woman (a legitimate target under his or her aggression) should die to avoid further bloodshed, as well to secure a quicker victory, then surely assassination through the just war theory is justifying the actions.

3

How Does God Feel about War?

So how does God feel about war, and does He understand? This question could go both ways; Scripture focuses that genuinely God supports and times He ordered the battle. You then turn over to other scriptures stating straightforward God rejects war. Well, if we look, we can find scriptures that have God telling King Saul to go and completely destroy the Amalekites and all they had and not to spare anything (1 Sam. 15). In verse 3, he stated, "Spare them not; but slay both man and woman, infant and suckling, ox and sheep, camel and ass." When Saul did not do exactly as God was telling him, God sent Samuel back to meet him. Saul told Samuel, "I have performed the commandment of the Lord," to which Samuel replied, "What meaneth then this bleating of the sheep in mine ears, and the lowing of the oxen which I hear?"

As a nation, whenever we are engaging in war, many antiwar protesters make their voices heard. Many demonstrators have started carrying signs with the slogan "Who would Jesus bomb?" Joining these protesters are religious leaders who state that war is inconsistent with the nature of God and the person of Jesus Christ. One of these protesters said, "Nothing I understand about Jesus Christ leads me to believe that support of war and violence are necessary or tolerable actions for Christian people." I believe that as Christians we need to know what God says about war. Didn't Christ come preaching peace, and if so, can war be compatible with life and teachings of Jesus Christ? Isn't war a contradiction to the teachings of God's Word? Let's turn to God's Word for the answers.

How can God be good if He forbade His people to protect their wives from ravishment and strangulation by drunken marauders?[1] Even to resist invaders who come to pick up their children and slaughter and abuse them? No policy would give freer rein to wickedness and crime than surrender for the right of self-defense for the law-abiding members of society. No more effective way of promoting the cause of Satan and the powers of hell to devise than depriving law-abiding citizens the right of self-defense. It is hard to imagine any deity could be "good" who ordains a policy of supine surrender to evil as recommended by pacifism.

Removing all possibilities of ordered societies would be abolishing any police force. Nations holding its liberty or preserving the lives of citizens must have an active defensive force available for defending against criminal actions. It is the perfect work of an excellent God to include the right of self-defense as the privilege of His chosen people. He would not be pleasant at all if He turned the world over to the horrors of unbridled cruelty perpetrated by violent and bloody criminals or the unchecked aggression of invading armies.

Christians have a duty and responsibility thinking through morality of the use of deadly force, especially when it comes to the issue of war. Many Christians are strict pacifists and lean heavily of the words of Jesus in the Sermon on the Mount:

> Ye have heard that it hath been said, An eye for an eye, and a tooth for a tooth: But I say unto you, That ye resist not evil: but whosoever shall smite thee on thy right cheek, turn to him the other also. And if any man will sue thee at the law, and take away thy coat, let him have thy cloak also. And whosoever shall compel thee to go a mile, go with him twain. (Matt. 5:38–41)

Does not it sound like a command for pacifism? Jesus also (in the same sermon) affirms the law written in the Old Testament,

> Think not that I am come to destroy the law, or the prophets: I am not come to destroy, but to fulfil. For verily I say unto you, Till heaven and earth pass, one jot or one tittle shall in no wise pass from the law, till all be fulfilled. Whosoever therefore shall break one of these least commandments, and shall teach men so, he shall be called the least in the kingdom of heaven: but whosoever shall do and teach them, the same shall be called great in the kingdom of heaven. For I say unto you, That except your righteousness shall exceed the righteousness of the scribes and Pharisees, ye shall in no case enter into the kingdom of heaven. (Matt. 5:17–20)

Does not it sound like a command for pacifism? Jesus also (in the same sermon) affirms the law written in the Old Testament,

Christian pacifists then point to a foundational teaching of the Old Testament, citing the command, "You shall not kill," found in the Ten Commandments. How can we, as Christians, justify the use of deadly force, even in times of war? Does not action of this nature violate the fundamental teaching of Jesus and the Old Testament (which Jesus affirmed as well)? Let's begin with a careful reading of the Scripture. "*You shall not kill*" is not a command found in the Bible. The command from scripture in the original language says, "You shall not murder" (Exod. 20:13). The Hebrew word for *murder* literally means "the intentional, planned killing of another person with malice."

Everything started tailspinning at the exact minute that Adam and Eve sinned against God and gave in to the temptation of the serpent.

> For God doth know that in the day ye eat thereof, then your eyes shall be opened, and ye shall be as gods, knowing good and evil. And when the woman saw that the tree was good for food, and that it was pleasant to the eyes, and a tree to be desired to make one wise, she took of the fruit thereof, and did eat, and gave also unto her husband with her; and he did eat. (Gen. 3:5)

Seeing this example, we arrive at one conclusion. Starting from beginning of time, Satan has controlled people nonstop, starting with a simple apple. Focus and think, Adam and Eve thought they would receive more than the gifts and blessings they already owned by eating the fruit. A feud develops, and, sadly, before you know it, there is shedding of blood. God is a God of peace, not war; He can defend His people; however, we must be in harmony with His grace to receive His help. The children of Israel while walking through the wilderness had hard times because they just would not listen to God's voice.

God spoke to the children of Israel, and it was at the exact moment they stopped listening that they ended in battle. God was trying to teach them they had to be dependent on Him, most likely the reason that we have conflicts in the world today. God just wants all His people to know that He is the Supreme Being of all, and we can fall back and know that He is going to catch us. God will not allow us to hit the ground, but if He has to, He will allow trials to happen so He can get your attention.

When reading the Word, one sees "eye for an eye." You flip the page, only to find another passage that says, "*Not thou shalt kill.*" It all just makes one ask, "So which way do I go in this?" God has shown himself as being a loving God that loves us all; however, it also shows us a vengeful God that gives a battle a sure victory. Ecclesiastes tells us, "To everything there is a season, a time for every purpose under heaven." He goes on to say, "A time of war and a time of peace." War is part of life, and God does understand this.

Philistine and Israelite armies were raging against one another in the battle array on opposite hills nonstop. The story of David and Goliath is a prime example. David, in the name of the Lord, went and fought Goliath and defeated him and then cut off the Philistine's head. The power from God enabled David's victory over Goliath; being supreme, God made this possible. David used a sling and rock backed by power from God. It was not David who went into battle; David allowed God to use his hands to defeat the giant.

Okay, so David killed Goliath. Was that wrong? Did he sin against God? Was God still encouraging? Let's look and see all through the Bible if we read we see how God was always ready for battle when it came to the Jews, when they were doing what they needed and living exact with God. With David thinking with the right mind setting, he knew God was the Almighty One, and through faith, he stepped on the battlefield. David was present in the body, but God was working through David to slay Goliath. In every reading, you find God always supplied victory when people obeyed Him. People cannot jump up and decide to take the life of people, you crossed the fine line. The difference is in war. We defend other nations needing support or our own at any cost. You need to focus and keep clean with your spirit before God, our maker and giver of life.

Need proof God can and has approved war? Let's see, Exodus 15:3 tells us, "The Lord is a man of war." Look over in Psalm 18:34 and Psalm 144:1 and hear how "He teacheth my hands to war." We see that not only God is a man of war, but He also teaches people's hands to war. That says that God the Father does all training, and when He trains us, how can we lose if we stay humbled to Him?

Let's look at a few references about God approving of war. We start in Genesis 14:1–20, where the Jews won the battle at Sodom and Gomorrah. Abraham gave a tithe to Melchizedek, and Melchizedek gave victory to God after Abraham went to get his brother's son being captive.

Abraham did not think twice about the mission. He was just in tune with God the Father, that his reaction was instant. Same with a soldier. We know by muscle memory what we need to do at what time, and we now do it. God leads people if they listen to Him and open their hearts, no matter their location. Your instinct tells you if it is enemy or friendly and then fixes to remove any hostilities through Christ our King.

Another story we find in Numbers 21:21–35, where the Lord spoke to Moses and said, "Fear him not: for I have delivered him into thy hand, and all his people, and his land." God was not only approving them to go into battle but also telling them that He will hand Og the king of Bashan into his hands and all his armies. We can arrive at a conclusion at this point God is a God of peace; however, He can make war. The oppressor is in control; he can stop before full conflict hits or continue until victory or defeat.

We find also that God commands Joshua to attack and gave strategic commands against Ai.

God knows and understands war will happen; however, God will not condemn you if God is first in your life and you follow God's plan. Our Heavenly Father must be the commander, seeking God not only during conflicts and war times, but times that we are peaceful and calm. Understand when we fight in battlefields, we must follow commanding orders from our leaders; however, try during time alone, bending your knees before God and seeking for the right orders. We can achieve a higher standard for other nations to follow. We must allow our prayers to reach God so He can work through our leaders for a perfect conclusion. The main question is, Can some battles be more victorious by simply putting God ahead of the battle? Yes, we have to go into battle when the country asks us to, but we need to understand that God will not judge against us for battling against one's enemies.

Sitting on point in the desert, the jungle, city, wherever a battle may be, take a battle position get ready for the enemy and let God work. We need to realize God finds you innocent when taking the life away for enemy soldiers in combat, but make sure you have listened to God and what He says. Understand you are not planning the killing; you are eliminating the enemy defending another country. No one likes the thought of taking another life, but, yes, there are times that war takes place, and we, have to live out whatever we have to do without guilt of God the Father casting all care from us. Let's understand God allows war to take place, and knowing He has defended for believers; however, He is loving and merciful and wants people to live a long peaceful live together in harmony.

Everyone has to take a stand, and if that means defending the vast nation we live in, in a combat zone, then stand tallest. Do not allow doubt from Satan to destroy your mission and gain the victory from you. Losing battles might happen continuously, but the victory falls with God and the people who follow. Yes, events happen that sometimes do not feel right, but the Lord is a Man of war and does understand. It does not make your steps easier, but know that God can and will help you on anything you need.

You're fighting the demons from combat because of the actions you performed on the battlefield and lives taken by your weapon? Not sleeping because you relive the images you have seen and faces you watch as they take their final breath of life? Sleep peacefully knowing God is merciful and waits for us to fall on our knees before our King.

People express their views and thoughts about combat and how they feel; however, you don't know until you walk there. People voice thoughts of who they want as president, who they want to work beside them. God gives us the right to make one's own decisions and to make them in the right way. Do we always do this? No, we do not. We have various faults, and God forgives us as long as we accept it.

God wants freedom to ring and ring loud; however, He also expects us to decide what makes sense when we are facing them. Do I go to war with my neighboring country because of them not being like me? No, God does not want us to battle because we cannot see eye to eye. God wants peace on the world He created; however, we won't allow God to work through us to follow His direction and the result war.

Our Father in heaven, He understands and knows the evil that is on this planet. Because of the evil, he allows us to defend ourselves, home, country, nation—whatever is under threat. Getting closer to home, we have police officers that defend our towns, city, and states from the evils of this world. I feel we underestimate God at times. We must remember God is a fair and just God.

Even with God, war was the extreme last action He approved of. God would give warning after warning before He would roll up His sleeves, wash His hands, and sanction the use of force against any nation. The rules of war laid down in Deuteronomy 20 represented a control of justice, fairness, and kindness. When using the sword, they did reflect the goodness of God.

> For there fell down many slain, because the war was of God. And they dwelt in their steads until the captivity. (1 Chron. 5:22)

When you have God siding with you, nothing can stop you. Defeating Satan only comes by using channels through God. Constantine the Great is a prime example. He received a vision that told him if he held the sign of the cross up high as they were going into the Milvian Bridge battle and become a Christian after the victory, God would see them through the battle. What do we think happened? Yep, he did just as God had asked, and winning the battle began.

We can also look at Gideon. God told him to gather men from the tribes of Asher, Zebulun, and Naphtali, as well as his own tribe Manasseh. By doing this action, God commanded they would be able to have an armed force of people from two tribes. They crossed the Jordan River, encamped at the

Valley of Jezreel. Listening and trusting God pays off. God told Gideon after he had gathered the men together that he gathered too many. God knew Israel would try to claim victory for them and not give credit to God for victory.

God first directed Gideon to send home those men who were afraid. Gideon then allowed any man who wanted to leave, to leave; twenty-two thousand men returned home, and ten thousand remained. God still knew he had too many. During the night, God commanded Gideon to approach the Midianite camp. Gideon overheard a Midianite man to tell a friend of a dream in which God had given the Midianites over to Gideon.

Gideon worshiped God for His encouragement and revelation. Gideon returned to the Israelite camp and gave each of his men a trumpet and a clay jar with a torch hidden inside. Marching into the enemy camp, Gideon and the three hundred divided into three. We have here proof that God will fight the battle for us when we put all faith and trust, blended with obedience, to Him. Furthermore, we can arrive at the conclusion that God, even though He does not like war, He will approve and aid in the battle. We see now God has agreed and helped in battle. God knows well that battle causes death and that the enemy placed themselves in that position. We must ensure all efforts to try settling the friction did take place before the fight.

So how can God love me and be in favor of me when I am a soldier going into battle and taking lives? Stop and realize you are a soldier, living a life of death and destruction; you are preserving peace and friction for people not able. I know our mortal minds cannot understand these issues; however, we need to know being a soldier for the military and living for God together can be performed. Your actions will distinguish you as a follower of Christ. God is a mighty God, and He is full of love and understands we are going to have wars, and we must protect the land, families, and home. God understands that in protecting territories from evil, not only are lives taken but bloodshed will also take place.

To help put minds at ease that God still loves you and does not judge against your actions in combat, then seek answers. Take this scripture as an example. Ecclesiastes 3:1 tells us "To every thing there is a season, and a time to every purpose under the heaven:" He goes on to say, "A time of war, and a time of peace."

God allows war to punish the wicked and protect the innocent.[2] He also allows war to show us our failures and guide us to Him. Where does that leave us? That leaves us on our knees.

> I exhort therefore, that, first of all, supplications, prayers, intercessions, and giving of thanks, be made for all men; For kings, and for all that are in authority; that we may lead a quiet and peaceable life in all godliness and honesty…I will therefore that men pray every where, lifting up holy hands, without wrath and doubting. (1 Tim. 2:1–2, 8)

If ever we need to trust the promise of Romans 8:28, it is in times like these:

> And we know that all things work together for good to them that love God, to them who are the called according to his purpose. (Rom. 8:28)

1. Although war is always dreadful and never God's ideal solution, war has been God's idea.
2. War is divinely delegated to the government, God's ministers who are called to protect the innocent and punish the evil.
3. A moral war is limited, not universal; national, not personal; defensive, not aggressive.

Our role as Christians during times of war is on our knees seeking God in prayer.

> For we wrestle not against flesh and blood, but against principalities, against powers, against the rulers of the darkness of this world, against spiritual wickedness in high places. (Eph. 6:12)

Yes, we can stay faithful to our military and live for God in harmony together without guilt for battlefield casualties inflicted at your hand. God does not punish those who are fighting to defend their country. I have heard some make statements like God will not love me because I have taken an enemy's life during a battle, but the truth of the problem is Satan wants us to believe that so he can win.

God protected and fought for the Jews through their hard times when they followed God's plan. He allowed hardships and pain from sin to fall on them to show they were slipping away from His will. We must start recognizing that God is the force behind the fight. We need Him from the start, and focusing on God will give complete victory. Even if your command does not want to give God the glory for actions, we can still do the act of giving Him praise. You must start with the individual first, then God works through us to change people from there. God will not and does not judge people accountable for what other individuals do; He deals with that person.

God strengthened individuals for war, including Moses, Joshua, and many of the Old Testament judges who proved great faith in battle. And God destroyed many armies challenging the Israelites. In 1 Chronicles 14:15, it describes God striking down the Philistines. God even gives counsel to be wise in war.

> Every purpose is established by counsel: and with good advice make war. (Prov. 20:18)

Today, America continues to face the horrible realities of our fallen world. Beaming suicide bombings and terrorist actions live into our homes daily. This serves as a constant reminder continuously the frailty of our flesh.

Obviously we must defend God-authored freedoms.[3] Throughout the book of Judges, God calls the Israelites to go to war against the Midianites and Philistines. This is because these nations were trying to conquer Israel, calling God's people to defend them. President Bush declared war in Iraq to defend innocent people. In fact, Proverbs 21:15 tells us, "It is joy to the just to do judgment: but destruction shall be to the workers of iniquity."

4

Why Does God Allow War?

WE FIRST NEED to understand that war is not necessarily a sin; however, what causes war is from sin. As we discussed earlier, man allowing themselves to fall into the trap of sin causes war among us today. In every situation, man has allowed greed or hatred to build to the point where people will not listen to anyone trying to work conflicts out peacefully. James reminds us that the supreme cause of war is lust and need. This restlessness that is part of us results from sin, this craving for that which is illicit, which we cannot stand. It shows itself many ways, both in our personal, individual life and also in the life of nations. It is the cause of theft and robbery, jealousy and envy, pride and hate, and infidelity and divorce. In precisely the same way, it leads to personal quarrels and strife, also to wars between nations.

Wars are the ultimate manifestation of man's selfishness, greed, pride, and competitive spirit and have caused untold suffering throughout history. The Bible says, "*From whence come wars and fighting's among you? Come they not hence, even of your lusts that war in your members?*" (James 4:1). People are to blame for the suffering caused by war. Their selfishness, greed, pride, and competitive spirit cause the destruction of others.[1]

The people of Europe looked at a question during World War II. They saw the suffering, death, and destruction of war and saw war was pointless, and they could not come up with an answer about why someone was allowing the war to happen. It is one expression of sin, one result of sin. When we look at any larger, more terrible occurrence, what is behind it is always sin.

Clearly, God allows war in order that men may bear the effects of their sins as punishment. This is a fundamental law that expresses itself in such words as, "Whatsoever a man soweth that shall he also reap" (Gal. 6:7). Punishment is if we do not change supremely in the next world for eternity; however, if one changes their ways before Christ's return, He spares them from the final punishment. God does allow individuals to suffer punishment in the foreground now if they do sin according to His Word.

God dislikes war, but to put an end to war, He would have to put an end to freedom of choice. He will one day soon, but for the moment, events have to play out. But it won't always be like this. The day is coming when Jesus Christ will return to put an end to the senseless suffering man inflicts on his fellow man. Then and only then, under the all-powerful rule and reign of Christ, will there be peace and plenty for all, with no more suffering, no more hunger, no more starvation, no more poor, and no more war![2]

Some might seriously ask the hard question, "But why do the innocent suffer?" You cannot arrive at an answer at this point; essence is twofold. First, there is no such person as the innocent, as we have seen already. We are all sinful; furthermore, we clearly reap the results not only of your personal sins but also the sins of the entire race and, on a smaller scale, the sins of our country or group, all people being the same in time—individuals and members of the state and entire race. God allows war so men see through it, more clearly than ever before, what sin is.

God withdrew His protecting care from them, and they were at the mercy of their enemies, who attacked them and robbed them. The kinds of statement that kept rising to the surface are God does not care about what is happening on the earth, He is too weak of a God to do anything about it, or He does not exist at all! Just consider the children of Israel again when they were not committing any sins against God. They had a sizable path to walk down on. God was always taking care of them, especially when he delivered them from bondage.

All this, in turn, leads to the final purpose, which leads us back to God. No word is found more frequently in the Old Testament as a description of the children of Israel than the words *"in their trouble and distresses they cried out unto the Lord"* (Ps. 107:6), all people still being the same. Indeed, as I contemplate human nature and human life, what astonishes me is not that God allows and permits war but the patience and the long-suffering of God.

Oh, the amazing patience of God with this sinful world! How wondrous is his love? He has sent the Son of his love to our world to die for us and to save us. Because men cannot and will not see this, He permits and allows such things as war to chastise and punish us, to teach us, to convict us of our sins, and, above all, to call us to repentance and acceptance of His gracious offer.

We have competition; everything is competition in this world we live, carrying along on the selfish, getting the basis of greed and vanity. To prevent the evils of competition and wars today, God would have to cram his religion down people's throat. The individual's way, violating the law of God and the law of love, is causing war, human anguish, and human suffering. God had to allow it (to let us have our own way) to fulfill his purpose of creating a holy character.

Not having free moral force or any character and fulfilling God's plan could never be. God allows war; therefore, God allows suffering from wars. Preventing war would be preventing the mere purpose that God is working here below. Do people know that man learns by suffering? Did you ever learn the stove was hot by putting your finger on it? We do learn by experience. Do we know that Jesus Christ learned by experience?

> Though he were a Son, yet learned he obedience by the things which he suffered; And being made perfect, he became the author of eternal salvation unto all them that obey him. (Heb. 5:8–9)

So in the face of wars, how do we prepare people for questions such as, "Why does God allow innocent people to get caught up in these terrible issues"?

First, we remind people there is no such thing as an innocent person and an innocent world. God created the world from the beginning without blemish, and man committed sin against God's command. He made man in His image and made him perfect, without sin or fault. He made man a personal being, able to having a perfect relationship with God. He created man as a moral being, giving him a conscience so he will be able to tell the difference between right and wrong, honorable and faulty.

He created us as a rational being, able not merely to think or draw conclusions but also to make moral choices.

So in this perfect condition, God gave man the ability to obey Him or the liberty to disobey Him. Sin came into the world, and man disobeyed God, and these are the results, shattering man's relationship with God. Man still holds his spiritual nature but lost his spiritual life.

Second, he lost his innocence and his moral free will by becoming tied down with ungodly ideas.

Third, by ruining his own personality with God, he lost his self-esteem, became guilty, ashamed, embarrassed, afraid, and anxious.

Fourth, he poisoned his relationship with others by suspicion, and distrust, and, lastly, his body became subject to decay, disease, and death. There is a human condition man has inherited that, as Paul says, "All have sinned." Sin entered the world through one man, and death through sin, and in this way, death came to all men *"because all have sinned."*

Even though no one is innocent, there is such an experience as innocent suffering. The case of Job's goodness highlights this fact, directly relating most suffering to sin, as well suffering in the world not related to sinning. Undoubtedly, you can point indirect connections by appealing to other scriptures about the fall and universality of sin, but they do not rob the book of Job of the point. For example, when a baby is born and born with AIDS, her suffering is the result of sin, someone else's sin; she is innocent. Even an eighteen-month-old baby covered with cigarette burns inflicted by her father. It was the abusive father who did it; the eighteen-month child was innocent.[3]

Look at Job again. The losses he faced were the result of human making—the Sabeans and the Chaldeans who took all his livestock and servants and then the natural disasters, the fire and wind that killed all his children. Behind them stood Satan, and behind Satan stood God. These issues happened to a man who God calls blameless, upright, and God-fearing. The Bible insisted all sinners will suffer; however, there is no insinuation that each instance of suffering is retribution for sin. If our world were not a fallen world, it's doubtful no suffering would exist. Just because it is a fallen world does not mean there is no innocent suffering. Considering the root of all problems, it is the symptoms that still bother people. Of course, we must draw people back to the root of the problem, but how do we deal with the symptoms?

Why does God allow wars? God does allow war to show the folly of man. To quote L. Jones, "It is not that war is sin, but that war is an effect of sin, or if you prefer it, that war is one of the expressions of sin." L. Jones makes three main headings. He answers the objection, "Yes, but why do the innocent get caught up in wars?" We have to reap the outcomes, not only of one's own sins but also the sins of the entire human race. He continues, "We share the same sun and rain, as other people, and we are exposed to the same illnesses and diseases. We are subject to the unhappiness of war because of industrial depression trials. Thus, it comes to pass the innocent may have to bear part of punishment for sins for which they are not directly responsible."

Second, he says that God allows war so man may see through it what sin is, and, last, he says that God allows war with the purpose of bringing us back to God. Clearly, war then is as an act of God's judgment. That does not mean to say there is no such case as a just war. Indeed, the Second World War, in one sense, was just. It does not mean that you need to enter a war hastily without making every single effort for peace.

However, war never achieves its absolute nature because "war is never an isolated act;" "war does not consist of a single short blow;" and "in war the result is never final." Christ Himself announces that wars will persist until he comes again, "You will hear of war and rumors of war" (Matt. 24:6). The cause clearly for war, as Jones puts it, is lust and want. We need to remind people continuously of the original problem. As Albert Einstein says, "What terrifies us is not the explosive force of the atomic bomb but the power of the human heart, in its explosive power for evil."

Being a soldier of Christ is straightforward as well. We need to start getting humble before God and showing Him that we want to be in the ranks. Keep yourself healthy, trained to be ready to go in a moment's notice. "Anyone not prepared to give your life for God today?" David Wilkerson asked several years ago to the people who were working with him. If they answered no or hesitated, he would tell them to "go and pray some more" until they got to the place they were willing to die for Christ. No use worrying; it is foolish. God is the only one that can be in command for everything to work.

In the story of the children of Israel, we see it time and time again. They disobeyed God and flouted his holy laws. For a while, all was well, but then they began to suffer; God also allows war so men may see, more clearly than they ever have before, what sin is. We tend to think lightly of sin in times of peace and to hold optimistic views of human nature. War reveals man and the possibilities within man's nature. A time of crisis and war is no time for superficial generalizations and rosy, optimistic idealism. It forces us to examine the exact foundations of life. It makes us face the direct questions about what it is in human nature that leads us to such calamities.[4]

The explanation cannot be found in the actions of certain men only. It is something deep down in the heart of man, in the heart of all men. It is the selfishness, hatred, jealousy, envy, bitterness, and malice in the human heart that shows themselves in the personal and social relationships of life, displaying them on a national and international scale. To explain them away, they seem to be excused within a personal sphere, but on the larger scale, they become more obvious.

God is perfect, absolutely holy, and infinitely just and fair. He is full of love and mercy, and therefore He must have a very powerful reason for allowing things such as suffering to take place. If God cannot sin or fail, certainly there must be an unassailable reason. The important thing is to want to know the reason why God allows wars, as well as suffering.

Sometimes we do not want to know the reason. Alternatively, we just do not believe there is a reason. We tell ourselves that because wars and suffering occur; therefore, there cannot be a God Who cares, certainly not a God of love. But there is, and our thinking should proceed along these lines: "If a perfect, holy, powerful and loving God allows such things, then certainly there must be a reason of such magnitude, that they are completely unavoidable."

Mankind has seized this world from God and rebelled against Him. We have turned our backs upon God. Indeed, we have spat on Him; we have spurned and rejected God's government of us. We have rejected God's ownership of us, His companionship also, and we have thrown aside His help, protection, and love.

We have all abused Him, sinned against Him, fouled up His world, and found ourselves fall into all kinds of trouble. We cannot make mistakes work out, and so we suffer the penalties of our conduct, the forfeiture of God's protection and help. Our pride will not let us admit our folly, and we use the mess which we brought about (including war and suffering) as a means of disproving, there could be a God of love. You could laugh, however, if tragedy is ugly.

Imagine a violent mutiny on a cargo vessel somewhere in the English Channel. The crew murders the officers and throws the captain into the sea. However, they are incapable of navigation, and within twenty-four hours, they have fought among themselves, set light to the ship, and hit the rocks off Land's End. Would you expect them to blame the owners of the vessel? Or to conclude that because they were in this mess, the owners, captain, and officers had never existed?

Man has brought about all his own problems, and, far from proving that there is no God, wars, suffering, illness, and tragedy all serve to prove that teaching of the Bible, which says that man has rebelled against God and is now the victim of his own sinfulness.

Many people say, "I can't believe that. I can't believe that man is that bad." But just look at people when they are engaged in the savage annihilation of others. Nations will go to war when practically nothing is to be gained. People will kill for greed and for personal power.

In the last decade, countless lives have been lost in bitter and vicious wars, several of which continue even now. In one country, thousands of the people perish from starvation while civil war racks their land. The Bible teaches us the human race is so bad there would be total war nonstop; however, the restraining hand of God prevents this from taking place.

Look at the way people exploit each other, tread on each other, and hurt each other. Look at the way men and women will fight in marriage, irrespective of what happens to the children. Look at the horrible selfishness and deceitfulness of which we are all capable. People expect, to some extent, for the world to revolve around them, as though they are the only person who matters. So if I experience a tragedy or if I am a parent and one of my children has a terrible illness and perhaps dies, I cry out in bitterness, "Why does God allow this to happen to me?"

Sin is not merely what we do; it is the state in which we find ourselves. We sin because we are hopeless sinners by nature. We are rebels, and we are spiritually and morally fallen creatures. So we are corrupt, proud, selfish, self-centered, and highly deceitful. We are also malicious and sensual, and we have stolen our lives, our years, our gifts and abilities, and seized them for ourselves. God, as far as we are concerned, might as well be dead.

What then do wars, tragedies, and sicknesses show us? They show us that God is separated from us and will not bless us. They show us people are on our own. In fact, they show us that God must punish us for our sin. If God must punish us presently, what about the future when life comes to an end? When we draw our last breath and cross over into eternity, what will He do then? Clearly, if we have never turned to Him in repentance, He must sentence us to everlasting banishment from His presence, and we must bear the punishment that is due to us.

Here is the wonderful grace of God upon people. He is still ready to pardon and forgive all who turn to Him. God is a great God of love, but wars and sufferings from wars warn us that He cannot endure sin and that He must and will punish rebellion. They warn us to turn to Him while there is still time and discover His mercy and converting power.

So if God is all powerful, loving, and merciful, how could He allow the misery and horror of war? To find God's answers, we must see what He recorded for us in the Bible about why we have warfare today. James, the half brother of Jesus, wrote under inspiration, "From whence come wars and fightings among you? come they not hence, even of your lusts that war in your members? Ye lust, and have not: ye kill, and desire to have, and cannot obtain: ye fight and war, yet ye have not, because ye ask not" (James 4:1–2).

As we wrap up this chapter. we must understand that God does not cause war, but we living people on earth do. It only takes a cursory glance at the history books to reveal that the major players in the history of warfare have always been men. But then again, how many of us actually know a man who has started a war? I certainly don't, and the last time I looked, I was still a man, and I still haven't launched any major armed conflicts I have only fought in them.

The issue of why men start wars has been investigated in a series of experiments by Professor Mark van Vugt of the University of Kent. They reveal that conflict is part of male bonding. "We all know that males are more aggressive than females but with that aggression comes a lot of co-operation," Prof van Vugt said. While male cooperation lies at the heart of democracy and leadership and men work better in hierarchical groups than women, it is a double-edged sword.

Our history teachers taught us that World War I began after a gunman killed Archduke Franz Ferdinand in Sarajevo on June 28, 1914. The shooting acted as a trigger, metastasizing from a Balkan street corner into a continental crisis by releasing pent-up tension between rival blocs of great European powers: the Austro-Hungarian Empire and Germany on one side and France, Russia, and Great Britain on the other.

The name of the gunman was Gavrilo Princip, his first name meaning "Gabriel" in his mother tongue, Serbian. His mother had wanted to call him Spiro after her late brother, but the local priest intervened, saying the boy should be name after the archangel Gabriel.

Most historians of the causes of World War II agree that its seeds were sown at the end of World War I. Some scholars say that the terms of the treaty were unnecessarily harsh and led to mounting anger in Germany, in particular over subsequent decades, but the BBC says, "It would be a mistake to imagine that the Treaty of Versailles was the direct cause of World War 2."

In August of 1939, German forces were built up along the Polish border in preparation of an invasion. Europe was still haunted by memories of the brutality of the First World War, and consequently, the governments in the region were reluctant to challenge the aggressive Nazis with military force. Most of Europe had looked the other way as the Nazis annexed portions of neighboring countries, but the leaders of France and Britain knew that an outright invasion of Poland must not be ignored. They pledged to rise to Poland's defense if necessary, placing the world a breath away from its Second World War.

On June 25, 1950, the Korean War began when some seventy-five thousand soldiers from the North Korean People's Army poured across the 38th parallel, the boundary between the Soviet-backed Democratic People's Republic of Korea to the north and the pro–Western Republic of Korea to the south. This invasion was the first military action of the Cold War. By July, American troops had entered the war on South Korea's behalf. As far as American officials were concerned, it was a war against the forces of international communism itself. After some early back-and-forth across the 38th parallel, the fighting stalled and casualties mounted with nothing to show for them. Meanwhile, American officials worked anxiously to fashion some sort of armistice with the North Koreans. The alternative, they feared, would be a wider war with Russia and China—or even, as some warned, World War III. Finally, in July 1953, the Korean War came to an end. In all, some five million soldiers and civilians lost their lives during the war. The Korean peninsula is still divided today.

The Korean War began as a civil war between North and South Korea, but the conflict soon became international when, under US leadership, the United Nations joined to support South Korea and the People's Republic of China (PRC) entered to aid North Korea. The war left Korea divided and brought the Cold War to Asia.

Before World War II, Vietnam had been part of the French Empire. During the war, the country had been overrun by the Japanese. When the Japanese retreated, the people of Vietnam took the opportunity to establish their own government lead by Ho Chi Minh. However, after the end of the war, the Allies gave back South Vietnam to the French while the north was left in the hands of the noncommunist Chinese. The Nationalist Chinese treated the North Vietnamese very badly, and support for Ho Chi Minh grew. He had been removed from power at the end of the war. The Chinese pulled out of North Vietnam in 1946, and the party of Ho Chi Minh took over—the Viet Minh.

In October 1946, the French announced their intention of reclaiming the north, which meant that the Viet Minh would have to fight for it. The war started in November 1946, when the French bombarded the port of Haiphong and killed six thousand people. The French tried to win over the people of the north by offering them "independence." However, the people would not be allowed to do anything without French permission! A new leader of the country was appointed called Bao Dai. The Russians and Eastern Europe refused to recognize his rule. They claimed that Ho Chi Minh was the real ruler of Vietnam.

On December 20, 1989, the United States broke both international law and its own government policies by invading Panama in order to bring its President Manuel Noriega to justice for drug trafficking. Noriega had seized control of his country back in 1983 when he became head of the national guard. From this position of power, he was able to build up the military and manipulate elections so that the winning presidents would be his puppet leaders.

Corruption was widespread during Noriega's rule, and he was able to use his power to imprison and sometimes kill any who opposed him. In 1987 a former officer of the Panamanian defense force publicly accused Noriega of cooperating with Colombian drug producers.

The causes of the Gulf War actually started when Iraq was at war with Iran. During this war, Iran was not only attacking Iraq but also attacking oil tankers from Kuwait at sea too. To support the ending of the war, Kuwait financially aided Iraq by lending the country US$14 billion. Iraq tried to convince Kuwait to dissolve the debt as Iraq had done Kuwait a favor by being at war with Iran. Kuwait declined, and this caused a rift between the two countries. For a year, they tried to resolve the financial situation but to no avail.

It was with disbelief and shock that people around the world saw footage of the terrorist attacks in the United States on September 11, 2001, when the planes-turned-missiles slammed into the World Trade Center towers and damaged the Pentagon. This ultimately resulted in the US declaring and waging a war on "terror." Osama bin Laden was eventually tracked down and killed some ten years later. But the way the war on terror has been conducted has led to many voicing concerns about the impact on civil liberties, the cost of the additional security-focused changes, the implications of the invasions and wars in Iraq and Afghanistan, and more.

So as you can see, nowhere in these wars—going back to the World War I to the current War on Terrorism—can you find and proof that God started them. Wars always begin when man gets greedy or just wants to control people. I always tell people, and this is the great thing, God could make us stop fighting. He has the power to make us serve Him as well. I believe we will all agree that God can do all things. Here is the great part: God loves us so much he will not make us but instead allows us to decide.

5

Gearing Up

EACH SOLDIER IN the Army wears what we always called battle rattle. Full battle rattle is close to fifty pounds worth of gear, including a flak vest, Kevlar helmet, gas mask, ammunition, weapons, and other key military equipment. One-piece is the light vest that covers the torso, shoulders, and back, making of the shell being of light material, a mixture of Kevlar and Twaron, then sewing them together in a sandwich fashion inside a nylon camouflage-pattern shell. The nylon vest has attaching points for load-bearing equipment. The second part are ceramic plates fitting in pockets, in the front, and back of the vest. These plates protect the heart and lungs.

Any TV news report from Iraq or Afghanistan shows American service members wearing full battle rattle. Wearing the battle rattle has saved lives in both Iraq and Afghanistan. A soldier in full dress, including helmet, flak jacket, and automatic weapon, is said to be wearing battle rattle. The terms *play clothes* or *Mommy's comforts* also are terms that antedated the war in Iraq, though used less often because they were using the gear by smaller numbers of troops. The term *battle rattle* was once associated with a call to arms on warships in the 1812.

A soldier must ensure that each unit of his or her gear is in tip-top shape and usable before going outside the wire. Each item is as critical as the other, combining all keeps soldiers safe; however, they must at all time remain alert and attentive to the opposing force. Their weapon must remain clean, and they must ensure that when firing at enemy personnel, the weapon does not jam and leave one defenseless.

When a soldier is sent to war, they go through combat proficiency training to prepare them for the mission. They are given immunization for diseases that they might face in combat and a description of the mission. They are prepared before they go to war. They must be mentally and physically ready for the task ahead. Preparing Christians for what they face in combat on the battlefield for Christ is like preparing the American soldier. Preparing ourselves physically, mentally, and spiritually for the battle that we will face is a must. We are going to war; our church has experienced a spirit of revival, changing people's lives, needing people to step up today as a leader in this fight God has.

Future Force Warrior is the Army's flagship science and technology[1] initiative to develop and display revolutionary abilities for the Future Force Soldier and Small Team. Employing humancentric systems approach for supporting the changing Army into a soldier-centrifugal force, the Future Force Warrior is a major pillar of the Future Force Strategy, complementing the Future Combat System and other Future Force programs.

Future Force Warrior links with advanced lightweight weapons and fire control optimized for urban combat, synchronized direct fire and indirect fire within and across Future Force Warrior team. They are creating lightweight, low-bulk, multifunctional, full-spectrum protective combat ensembles for better protection of our soldiers incorporating ballistic protection, novel signature management, semipermeable membrane for CB/wet protection, electro-textile power and data body LAN. Also included are on-board physiological and medical sensor suite with strengthened casualty care plus customized voice, tactile, visual, and auditory human link, with integrated laser eye protection.

To better equip its soldiers, the US Army is developing an advanced infantry uniform that will provide superhuman strength and greater ballistic protection than any uniform so far. Using wide-area networking and onboard computers, soldiers become more aware of action around them and of their own bodies. Developing a bionic uniform for its soldiers, the United States Army is planning a change in logistics of war. Integrated physiological overseeing, increased communication, and augmented physical strength will give the soldiers of the future the tools they need to overwhelm their opponents simply by donning a high-tech suit.

There were two phases to the Future Force Warrior program. The first phase involves placing of a uniform in 2010 that would meet the Army's short-term needs, although pieces of the uniform may be used earlier. According to Future Force Warrior Equipment Specialist,

"The Department of the Army has built what's called design spirals. This means roughly every two years, if a piece of technology has matured, we try to get it in the field, rather than waiting until 2010, to field the entire system." In 2020, the US Army plans on rolling out a suit that integrates nanotechnology, exoskeletons, and liquid body armor, all of which exist only in idea now.

Here are the basic units of the final version of the suit:

- Helmet—The helmet houses a global positioning system receiver, radio, and the wide- and local-area network connections.

- Warrior physiological status monitoring system—This layer of the suit is the closest to the body and contains sensors that monitor physiological indicators, such as heart rate, blood pressure, and hydration. The suit relays the information to medics and field commanders.

- Liquid body armor—This liquid body armor is made from magneto rheological fluid, a fluid that remains in a liquid state of the application of a magnetic field. When an electrical pulse is applied, the armor transitions from a soft state to a rigid state in thousandths of a second.

- Exoskeleton—The exoskeleton is made of lightweight, composite devices that attach to the legs and augment the soldier's strength.

Together, these subsystems combine to create a uniform that tells, protects, and strengthens the abilities of its wearer. Now let's take each item of the suit separately.

Another vital unit of battle is military communication between soldiers. The Future Force Warrior will use sensors that measure vibrations of the cranial cavity, cutting out the need for an external microphone. Bone-conduction technology allows soldiers to communicate with one another and also controls the menus visible through the drop-down eyepiece.

The helmet has 360-degree situational awareness and voice expansion. "What this will allow you to do is to know where that sniper round or mortar round came from. At the same time, it will cancel noise at a certain decibel not to cause damage to the soldier's ears," said a liaison sergeant, operational forces interface group, Natick Soldier Center.

The situation-awareness technology also allows soldiers to:

- detect soldiers in front of them up to a couple of kilometers away and
- focus in on a particular sound and expand it.

With advances in ballistics, armies must develop better body armor. One type of modern body armor, first developed in the 1960s, is made of advanced woven fibers that can be sewn into vests and other soft clothing. More commonly known as DuPont Kevlar, this is one of the many-body armor solutions currently employed by US Forces. Another armor for smaller arm's protection insert plates, or "small arms protective insert" plates, hardening ceramic composite plates inserted into a soldier's fragmentation protective vest in both the forward and back upper-body pockets.

In the shoulder of the Future Force Warrior uniform is a fabric filled with nanomachines that mimic human muscles, flexing open and shut when stimulated by an electrical pulse. These nanomachines will create to lift the way muscles do and augment the overall lifting ability by 25 to 35 percent. "Think of yourself on steroids, holding as much weight as you want for as long as you want," said Atkinson. "It will also allow a 90 pound male or female to carry a 250 pound male or female off the battlefield, and it wouldn't feel like they were carrying 250 pounds worth of a person."

"The Exoskeleton, which is with Defense Advanced Research Projects Agency,[2] will give the soldier more stability," Atkinson said. "It makes the soldier become a weapon's platform."

With this added strength, mounting weapons directly to the uniform system. In the concept uniform, the exoskeleton is the protruding composite material below the knee.

The exoskeleton will merge structure, power, control, actuation, and biomechanics. Here's a look at some of the challenges that Defense Advanced Research Projects Agency has outlined:

- Structural materials—The exoskeleton will have to be made of composite materials that are strong, lightweight, and flexible.

- Power source—The exoskeleton must have enough power to run for at least twenty-four hours before refueling.

- Control—Controls for the machine must be seamless. Users must be able to function normally while wearing the device.

- Actuation—The machine must be able to move smoothly so it's not too awkward for the wearer. Actuators must be quiet and efficient.

- Biomechanics—Exoskeletons must be able to shift from side to side and front to back, just as a person would move in battle. Developers will have to design the frame with humanlike joints.

Soldiers for God, like American soldiers, have protective wear aiding in fighting the evils of the world. Soldiers of God have the full battle rattle available to them; however, this battle rattle being lighter than that of the military soldiers. In the book of Ephesians, Paul writes his letter to the church at Ephesus, giving them instructions in his absence. Within this letter, Paul commands the church to put on the full armor of God. Just what did Paul mean by such a command? The war we are fighting is intensely real, not a metaphorical exercise; the war we are fighting takes place in the spiritual realms all around us. Throughout the Scripture, highlighting this point repeatedly, we see Elisha surrounded and protected by the army of the Lord as the Syrian army tries to take him prisoner.

David typifies this relationship between God's people and their role as God's army. God used David a shepherd boy who went in the power of the Lord as an example for all of us. David went to visit his brothers, who were parts of Israel's army; they were facing the pagan Philistines and their champion Goliath. Day after day, for forty days, Goliath would taunt the armies of Israel, daring someone to challenge him. No one in the army of Israel had the courage to fight such a tremendous specimen of humanity. Goliath was a soldier from his youth, and stood nine feet nine inches tall.

God's armor brings victory because it is far more than a protective covering, testing the exact life of Jesus Christ Himself fully through His death and resurrection. "Put on the armor," wrote Paul in his letter to the Romans, "Clothe yourselves with the Lord Jesus Christ" (Rom. 13:12–14). When one does, He becomes a hiding place and your shelter in the storm, just as He was to David. Hidden in Him, we can count on His victory, for He not only covers us as a shield, He also fills you with His life. Since living in the safety of the armor means oneness with Jesus, we can expect to share His struggles as well, as His peace. Remember, God offers us His victory among trouble, not the absence of pain.

Ephesians 6:12 clearly points out that conflict with Satan is divine; therefore; employing no visible weapon effectively can work against him and his minions. It does not give us a list of tactics Satan will use; however, the passage is clear that when we follow all the instructions faithfully, we will be able to stand, and we will have victory regardless of Satan's strategy.

Any other strength proves powerless against Satan. Your own strength never stands enough to oppose Satan, but even a little of God's strength is enough to win any battle. Paul said, "I can do all things through Him who strengthens me" (Phil. 4:13). It is not the strength we have that is prestigious only its source. The extent that Christians are strong in the Lord, His victory is guaranteeing us to overcome the worst Satan has to offer. We are in a state of war, a terrible and fierce war, but we have no the reason to be afraid if we are on the Lord's side.

Appropriation of that strength comes through what the Puritans referred to as "the means of grace" prayer, knowing Scripture, obeying it, and putting faith in the promises of God. To take advantage of the strength of God's might, believers must put His armor on.

> Put on the whole armour of God, that ye may be able
> to stand against the wiles of the devil. (Eph. 6:11)

The Greek word translated "put on" (*enduo*) carries the idea of permanence. Putting on the whole armor of God and leaving it on always, not occasionally but permanently, protects continually. When used in a military sense, the Greek word translated "stand firm" (*histemi*) refers to holding a critical position while under attack. Living obedient is what enables us to stand firmly in a Spirit-empowered life.

The Christian is in a no-man's land between the opposing forces of Light and darkness between the Almighty God and His ominous enemy, Satan, otherwise known as the devil. On your own, you cannot face demonic schemes and principalities, powers, and rulers of the darkness of this world. Therefore, we must align ourselves always with God and take the heavenly aid that God provides through His divine weapons of warfare and through His armor, the armor which the Son of God (Christ) Himself wears.

No soldier in his right mind would ever enter a battlefield with no weapon and without his protective armor, yet many Christians do this every day. Utterly defenseless, we are easy targets. Christians must begin their day by putting on each piece of God's armor. The armor found in and provided by Jesus, "*Put ye on the Lord Jesus Christ*" (Rom. 13:14). A soldier in the military needs to have his or her weapon when going out onto the battlefield. The weapon defends and protects from enemy forces, so should soldiers for God! Even if the enemy is surrounding someone, remember never to lose your hope. I felt my world would have caved in while I was fighting on the battlefield—looking back now, those times were the darkest—if God had not shadowed me. People need to realize seeing shadows mean there is a light shining down.

Gearing up does not only stand for times we are at war, whether physical or spiritual. We must remember that soldiers always prepare for the mission at hand. Training needs maintained, not a fast overlay; training can go for a couple of months. One issue I do want to make clear, I can tell the difference between the hard-core soldier and the ones who have the attitude of "I honestly do not want to be here." When on the home front, the hard-core soldier takes pride in everything they do, including wearing their uniform. I know from experience the soldiers who were always late and did not wear the complete uniform were the soldiers I was disciplining all the time, trying to get them to step up. They never carry out tasks on their own; they were told and watched nonstop, even on their own time continuing mistakes.

When individuals went through military entrance processing station and arrived at their first duty station, where they were issued all the gear, which we called basic issue, some thought, *Why do I need all this mess?* You think you don't need all the items issued there, but in the combat zone, you see how important they are; the items work for the better cause of the mission. Spiritual realms work the same way. God one's Great Commander sets us up to succeed by allowing us to encounter trials to strengthen us. We must set people and ourselves up for success, not for failure.

You do not want to find, after getting on the battlefield, that your equipment is damaged and not usable; it will be too late. Same goes with your walk with God. We must trust Him to show us what's needed to make it in the end. I would much rather go through torment here than eternal agony in the lake of fire. As the soldier in this physical world, we clothe ourselves in distinguished uniforms. We make sure that everything looks right every time we wear it. We also need to wear the blanket of righteousness with pride to let the whole world know that we are different and more powerful than anything this ole world can throw at us.

When a soldier goes into battle, he lays aside everything that does not pertain to the fight ahead. He puts on his uniform, takes up his weapon—prayer, the sword of the Word—praises and worships God, and waits for the command from above.

> Put on whole armour of God, that ye may be able to stand against the wiles of the devil. (Eph. 6: 11)
>
> For the weapons of our warfare are not carnal, but mighty through God to the pulling down of strong holds. (2 Cor. 10:4)

He is trained. He is under authority and does not go running off to do his own task. A soldier in the field does not see the whole picture, and what he does see can seem strange and even wrong and counterproductive. If he trusts his commander and believes in the justness and wisdom of his cause, he will follow orders, even lay down his life. When Satan "comes in like a flood" (Isa. 59:19), raise a standard battle flag against him and stand. The battle is the Lord's (2 Chron. 20:15). He will deliver!

6

Fighting the Good Fight

AMERICA HAS BEEN fighting battles since 1775, the American Revolutionary War led by General Washington. This started the long of line of battles America was to face. Looking through each battle, one common issue stays. Anger starts war and is the result of a national entity wishing to improve the standard of living for its people. A second important cause we find is when there is a possible decrease in current standard of living and the nation fights to protect what it already has.

If a human's actions are beyond his or her control, then the cause of war is irrelevant and unavoidable. On the other hand, if war is a product of human choice, then identifying three general groupings of action would consist of, biological, cultural, and reason. While exploring the cause of world conflicts, one must look into the relationship between human nature and war.

Finally, the question remains about whether war is ever morally justified. The just war theory we discussed earlier is a useful structure in examining which war may be ethical. In the planning background of modern warfare, a moral breakdown of war will need the philosopher of war account not only for military personnel and civilians but also for acceptable targets, strategies, and use of weapons. Eliminating enemy personnel is a taboo issue few soldiers have to do, and those who do prefer not to talk about it.

In waging war, it is considering unfair and unjust to attack indiscriminately since noncombatants or innocent is considering standing outside the field of war proper. Immunity from war can be from existence and activity not being a part of the essence of war, which is killing enemy combatants. Since killing itself is challenging, the just war theorist has to proffer a reason enemies become legitimate targets in the first place and whether their status alters if they are fighting a just or unjust war.

First, a theorist may hold that having training and/or being armed sets up a sufficient threat to combatants on the other side and that the donning of uniform alters the person's moral status to legitimate target. Whether this extends to times of peace is not certain. Voluntarism may invoke the boxing ring analogy. Punching another individual is not morally sustainable in a civilized community, but those who voluntarily enter the boxing ring renounce their right not to be hit.

Equally, in joining the army, they have been saying the individual has the need to renounce his or her rights not to be targeted in war. To bear arms takes a person into an alternative moral realm. Killing is the expectation and possible norm. It is removing the world from civilian structures, and history has devised rites of passage and exit that underline the change in status for cadets and veterans.

All analogies to the fair play of sports fail at this juncture for war involves killing and what the British Army calls "unlimited liability." On entering the army, civilians lose the right not to be targeted, yet does it follow that all who bear uniform are legitimate targets? You have some soldiers fighting, compared to those bearing arms who are involving themselves in supplies or administration, for instance. This principle overlaps the proportionality principle of just cause, being distinct enough to consider it in its own light. Jus in bello needs tempering the extent and violence of warfare to reduce devastation and casualties.

In recent years, the United States and the UK proclaimed the Gulf War was not with the Iraqi people but its leader and his regime. The US government even issued a bounty on the heads of key agents in the Ba'ath party—for example, Saddam Hussein's sons, Uday and Qusay—killing them in a selective hunt and destroying the mission rather than capturing and bringing them to trial for the crimes asserted of them. Assassination would clear the two hurdles of judgment, and yet the central wing of just war theorists would practically claim that underhand and covert missions, including assassination, should not form a part of war on grounds that they act to undermine the respect for of one's enemy (no matter how cruel he or she is).

Assassination would apparently clear the two hurdles of discrimination and proportionality, yet the constitutive wing of just war theorists would reasonably claim that underhand and covert operations, including assassination, should not form a part of war on grounds that they act to undermine the respect due one's enemy (not matter how cruel he or she is) as well as the moral integrity of the assassin; the consequentialists would also counter that such policies also encourage the enemy to retaliate in similar manner, and one of the sustaining conclusions of just war theory is that escalation or retaliatory measures (tit-for-tat policies) should be avoided for their destabilizing nature. Once initiated, assassination tends to become the norm of political affairs—indeed, civil politics would thus crumble into fearful and barbaric plots and conspiracies (as did Rome in its last centuries) in a race to gain power and mastery over others rather than to forge justifiable sovereignty. One of the continuing conclusions of just war theory is avoiding that increase or retaliatory measures (tit-for-tat policies) for their destabilizing nature.

Just stop and ask yourself about that feeling you got eliminating the enemy soldiers. Eliminating the enemy is a difficult act to do and is a sobering experience, happening extremely quickly, and when done, one is just happy they and their soldiers are still alive. Eliminating enemy forces causes some to question why the enemy soldier is so ready and willing to die for their cause. Are they convinced by religious and fanatical propaganda or a payoff to attack Americans? Did they think the full process out including the family and friends they are leaving behind?

Every soldier in my squad was decisively engaging themselves, pulling the trigger and taking the life of the enemy soldier or maneuvering to corner them around the town. We saw the enemy collapse, explode, and vaporize at one's own hands and civilians caught in the cross fire. As a Christian leader, what am I to do with these experiences? What can I share to help those leaders engaging combatively with the enemies?

There is a satisfaction in knowing that we got the job done well, destroying the enemy before they killed you or your friends and family. My experiences with taking the enemy's life while in combat haunted me for years. Knowing the battle zone that I would fight in, actions that was accomplished on the battlefield, changing the lives by my hands is hard to discuss. I am opening up a little each day. At first, I did not want anyone, including my family, to know what I had to do in combat. Taking a life causes one to evaluate your personal mortality and death. The issue of individual mortality is what I think bothers most soldiers and others who have taken a life in combat. They are afraid of death and what lies beyond.

The real challenge is life after deployment. Historically, those fighting in combat have a much greater likelihood of relationship breakups than their civilian counterparts. Depending on how devastating the war experience was, soldiers may be completely different from the person you saw leaving for war. He could have witnessed others, including children or comrades, die. He or she may have been forced to kill enemy soldiers in the line of duty! While your soldier may not share all these experiences with you, listen with empathy if he or she does.

To the soldiers out there, even though your life was clouded with fear during deployment, don't compete for the most wounded heart. After a big homecoming, you'll want to get on with your life and live normally but may find yourself and your spouse sitting at different junctures.

To the soldier's husband or wife, after returning home from a country at war, everyday life may now seem trivial to your spouse. He or she may suffer from postwar trauma or guilt.

How can a Christian, in good conscience, pick up a weapon and deliberately kill another human being? Human beings, by nature, are warmongers, and peace is a rare commodity, for how Christian perspective should be on this subject.

I thank God and praise Him every day for keeping me safe and providing us victory. I pray for all soldiers and their safety, but, most importantly, I pray for their salvation. I also pray for the enemy's salvation and for those greatly affected by their actions.

Today in modern times, we have four perspectives that deal with war. First, you have pacifism, which is someone against killing and against any war. This view says that calling Christians as peacemakers always excludes any participation in war any time for any reason in any capacity.

You have the nonresistant. Their view says Christians may take part in war but only in noncombatant roles—serving as a chaplain, a medic, driving a supply truck—any role that does not directly involve picking up a gun and shooting someone or firing of a missile and blowing somebody up.

We also find, as I've tried to explain, calling the just war view says that Christians may engage in what would be a "defensive" war to protect your country's freedoms and rights from outside aggression. Then finally, there is the preventative war. This view is an adjustment of the just war theory and says that Christians may go to war to stop an attack on someone else to or correct outrageous injustice.

Priority of training in units is to train to be standard on the wartime mission. Battle focus guides the planning, preparation, execution, and assessment of each organization's training program to ensure its members train as they are going to fight. Battle focus allows commanders and staffs at all ranks to structure a training program that copes with nonmission-related needs while focusing on mission vital training actions, recognizing units cannot reach proficiency to standard on every task, whether because of time or other resource constraints.

The term *battle focus* refers to an idea used in the United States Army to decide peacetime training requirements based on wartime missions. Most popular spiritual warfare books focus on the individual Christian's personal struggle with the weaknesses of his own flesh and his egocentric spiritual battle against demons and promises relief from the exhaustion of the conflict. That is in stark contrast to the training a soldier receives in the nation's armed forces.

Soldiers are trained to survive on a battlefield and carry out a mission. They are trained to defeat an enemy by fighting as a member of a team under the orders of you chain of command. The Lord of Hosts won everyone's eternal victory on the cross at Calvary. He gave His Church the mission to take the fight to the gates of hell and the Gospel to the ends of the earth regardless of personal costs!

Every aspect of modern culture has been engaging a spiritual battle with the eternal souls of all humanity. Christianity is not about merely securing a place in heaven after death but about redeeming this lifetime by yielding your lives to the work of Jesus Christ. Continuing in us and through us in this generation, one's enemy, which is the devil, and fights against us trying to over through our defenses. If we're living the Christian life, we are in a battle. God's people need faith and courage to fight the good fight of faith.

So for us to be in a spiritual warfare with evil, we need to be wearing the entire battle rattle. We talked about a little earlier on how the United States military use's equipment and how it relates to the Christian and their gear. We must keep our gear ready always so we can fight in a minute's notice. We as Christians are on an instant alert call. The military has units that do rapid deployment and have a short time line to follow. I was with a combat unit that had a seventeen-hour window to fly. With Christianity, we do not have that window. Satan will hit you without notice, and we must be ready for him.

Satan looks for the believer's weaknesses to use as ammunition against him. He finds a vulnerable area of your life (a difficult life status quo, a character weakness, or a past failure) and concentrates on that area, shooting the fiery darts of his temptations. If you disobey and repent, God forgives and restores, according to His promise to us in Scripture. If you disobey and continue to sin, Satan interprets that as an invitation to continue his actions, and a pattern begins, as Satan can dominate your mind and body more and more.

We do not fight against a frail human enemy but against spiritual forces of the devil. We do not go out and burn down buildings where one hears false teaching or tear down nightclubs, among other violent acts, but we use the sword of the Spirit (Eph. 6:17; Heb. 4:12) to conquer the hearts of men. The kingdom of Christ is not defended or extended either by carnal means or by a social gospel.

The highest art form in warfare is using the opponent's forces against him. Satan takes what God does and aligns with it and redirects it toward God while God takes what Satan does, aligns with it, and redirects it at Satan. The Bible says, "Do not be overcome by evil; but overcome evil with good." The wisest way to act is not forcing issues to happen, but with a natural flow of everything, and directing it to your advantage. Instead of rejecting the work of the enemy, we should use and change it.

Success is gaining in warfare by carefully holding ourselves to the enemy's purpose. Calling this ability to succeed is sheer cunning. Walk in the path defined by rule and accommodate yourself to the enemy until one can fight a decisive battle. Satan incorporates God's work into his plans, and God incorporates Satan's work into his plans. In the end, one plan will prevail upon God's plan.

A stronghold begins when we give Satan permission to enter our lives through disobedience; we allow him to have control. As the "god of this world," he has the legal right to influence and harass us if we open the door to him, and remember this important fact: God will, grant Satan's petition if no one contests it! If no believer (having Christ's authority over Satan) contests Satan's legal claim, God will grant Satan's request! God will not overrule a person's will if he chooses to disobey God allows man to disobey. That's why it is important to know who we are in Christ, your rights and privileges, so we can use our authority against Satan!

As "heirs of God and joint-heirs with Jesus Christ," we have more right to petition God than Satan does! That's why it is important to intercede in prayer for others—to "stand in the gap" for the loved ones Satan is blinding with strongholds that they can't recognize them or get free of. As believers, we can take authority over the enemy, using the authority of Jesus's name, and pray for them and contest Satan's influence. (You may be the only one who can cause a change in their lives) When we don't confront Satan by using our spiritual authority and weapons to destroy strongholds, Satan will continue to "steal, kill, and destroy."

If we know the enemy and know ourselves, we need not fear the result of a hundred battles. If we know ourselves but not the enemy, for every victory gained, we will also suffer a defeat. If you know neither the enemy nor yourself, you will succumb in every battle. If we know the enemy and know ourselves, your victory will not stand in doubt. If one knows heaven and knows earth, one may make your victory complete. We should seek to know God and Satan because one represents us, and the other is heaven's enemy.

I understand as we talk about fighting the good fight that it is easier to feel better about it being a good fight when we are referring to Satan. It is difficult to think about the procedure of killing any individual in a combat zone, regardless. The Bible provides a wealth of guidance on warfare and killing, with many examples of warriors serving the Lord, such as Joshua, Samson, and David. The Lord instructed Joshua to establish cities of refuge for those who were killed "*unintentionally and without malice aforethought*" (Josh. 20:5). That comforts me, having seen civilians fall in the cross fire of battle, knowing the Lord looks at your heart when evaluating your actions.

Don't you forget it! Satan is our deadly enemy, and he fully intends to destroy us. Cancel the idea that he is a comical imp dressed in red tights, sporting a tail, goatee, a pair of horns, and carries a pitchfork in his hands. On the contrary, he is a powerful spirit being who leads a host of highly organized fallen angels and demons.

Satan and his demons stalk the path of every believer, offering all manners of enticements to lure the Christian away from an obedient and faithful walk with Christ. No one is free from satanic attacks, and no one is completely successful in countering them (I John 1:8, 10), but some Christians succumb to temptation so often that they see no hope for victory. These Christians give up and give in to Satan's forces without a struggle. This is an unfortunate condition, born of despair, for it will blind the believer to the marvelous provision God has made for overcoming temptation.

No aspect of our spiritual warfare is more important than learning how to recognize and destroy them, for as long as they exist, they will cause frustration, hinder spiritual growth, and cause defeat in our spiritual lives. Satan's objective is to deceive us and cause us to become blind. This causes people not to see that they are there or become so discouraged with trying to overcome them that they give up and fall away from God, thinking, "It's no use. I'm just a failure at being a Christian!"

The Bible names him "Satan" (the adversary) and "the devil" (the slanderer). He is described as the dragon, the serpent, a roaring lion, a liar, murderer, and the wily, crafty accuser of the brethren. He is the epitome of all that is evil, and to top it off, he hates God and God's people. Jesus warned Peter, "And the Lord said, Simon, Simon, behold, Satan hath desired to have you, that he may sift you as wheat." (Luke 22:31).

Beware, Satan is our adversary.

> For we wrestle not against flesh and blood, but against principalities, against powers, against the rulers of the darkness of this world, against spiritual wickedness in high places. (Eph. 6:12)

He is called "the prince of this world." He is powerful, clever, vicious, and determined to destroy us. We must overcome him in our personal life. However, how can we ever expect to defeat him?

One of his most effective schemes involves division. He prompts one member to offend another and then urges the offended member to ignore God's instruction for dealing with offense. Satan leads the offended Christian to what comes naturally rather than what comes from the spirit. The body of Christ will be torn, and the church will lose much of its power. It's a simple plan, and it is tragically effective. When someone offends us, the inclination of our old, sinful nature is to hold that offense tightly. We want to linger on the injustice we've suffered. We need to tell others, longing for salve of their sympathy and encouraging their outrage. We crave eye-for-an-eye retribution, public shaming, and judgment.

We know Satan will attack along the way. We also know that we must be ready to fight at all times. There is great news! Victory is God's; victory is ours! God has wrought the victory already, but we still must be ready to fight.

7

Identifying the Enemy

AMERICA'S INTELLECTUAL FAILURE to identify the nature of the enemy is a major cause of its defeatism, but this failure, and its responsibility for our policies, only goes so far. For example, none of our politicians identify our enemy as "Islamic totalitarianism"; however, they all know and admit that Iran and Syria are active sponsors of terrorism, that Iran is developing missiles and a nuclear weapon, that Saudi Arabia turns out legions of wannabe terrorists, and many other facts pointing to the conclusion that if we are to be safe, these states must be stopped. Shortly after 9/11, President Bush demonstrated some understanding of the role of state support of terrorism when he declared, "From this day forward, any nation that continues to harbor or support terrorism will be regarded by the United States as a hostile regime." Recently, despite his misgivings about indicting any variant of any religion, he has been condemning "Islamic radicalism" as a major source of the terrorist threat.

If America were to take military action to end the threats we face, even based on our leaders' limited understanding of these threats, it would be far more significant and effective than what we have done so far. Why then haven't our leaders taken such actions?

The reason is that despite their claims that they will do whatever is necessary to defend America, our leaders believe that it would be wrong—morally wrong—to do so. They believe this because they consistently accept a certain moral theory of war—one that has come to be universally taught in our universities and war colleges. This theory is accepted, at least implicitly, not only by intellectuals, but by our politicians, the leadership of our military, and the media. And while the American people are not explicitly familiar with this theory, they regard the precepts on which it is based and the policies to which it leads as morally uncontroversial. The theory is called just war theory. To understand today's disastrous policies and reverse them, it is essential to understand what this theory holds.[1]

Of all the dramatic images emerging in the hours and days following the September 11 terrorist attacks, one of the most haunting images was a frame from a surveillance-camera video. This image captured the face of suspected hijacker Mohamed Atta as he passed through an airport metal detector in Portland, Maine. Even more chilling to many security experts was, if we had the right technology in place, an image like that might have helped avert the attacks by identifying him. According to experts, face recognition technology that's already commercially available could have instantly checked the image against photos of suspected terrorists on file with the Federal Bureau of Investigation and other authorities. Making a positive match, the system could have sounded off the alarm before the suspect boarded his flight.

In the wake of the attacks, several companies, security professionals, and government officials have proposed using biometrics identification based on a person's unique physical characteristics to strengthen airport security. "We've developed some fantastic technologies, but we just haven't deployed them," says Georgia State University aviation safety researcher Rick Charles. Readily available biometric techniques include fingerprinting digitally, scanning the iris of your eye, voice recognition, and face recognition.

For the terrorists, suicide bombing is a communal safety valve. In an outrageous spasm of perverse creativity, it serves to protect the broken Palestinian community from its own violence, which always is a great deal. Arab and Islamic terrorism, therefore, serves the critical interests of Palestinian social solidarity, although Palestinians could invent any other enemy to serve as a convenient deflection from Palestine's own continuous eruptions into strength. Selecting Jews for the victim role is indisputably most consistent with various theological expectations and prescriptions.

Two things distinguish the irregular wars in Iraq and Afghanistan: t is not clear who is a combatant, and the United States is fighting a conflict with no clear battlefront.

For much of the period following 9/11, the Bush administration claimed that no legal authority could constrain the executive, regardless of the source of law and regardless of whether at home or abroad or whether applied to enemy combatants or US citizens. In response, civil libertarian objectors tried to fashion claims that domestic law should apply to all uses of American national security power, adopting the same blanket approach as the administration, only in reverse. When pressed to the point of claiming that there must be full judicial proceedings in a war zone, as in the recent case claiming habeas corpus rights for prisoners at the Bagram air base in Afghanistan, the civil libertarian claim collapses as lacking legal authority and being thoroughly impractical.

Positive identification of friends and foes is a matter of life and death for US troops and local citizens in Iraq and Afghanistan. Mistake an enemy for a friend, and a terrorist gains access to a crowded US base or a secure town. Mistake a friend for an enemy, and a supporter is lost, or an insurgent is created.

The United States Department of Defense defines terrorism as "the calculated use of unlawful violence or threat of criminal violence to inculcate fear. Anything intended to pressure or to intimidate governments or societies in the pursuit of goals that are generally political, religious, or ideological." Within this description, there are three key parts; violence, fear, and intimidation, each element producing terror in its victims. The Federal Bureau of Investigation uses this, "Terrorism is the unlawful use of force and violence against anyone or property to intimidate or coerce a government, the civilian people, or any segment thereof, in furtherance of political or social intentions." The US Department of State defines *terrorism* to be "planned politically-motivated violence perpetrated against non-combatant targets by sub- national groups or clandestine agents, usually intended to influence an audience."

Palestinian terrorism, we see, is an intensely religious ritual of human sacrifice. Interestingly, in all such ritualistic killings, the victims must bear a basic likeness to the "sacrificers." At the same time, they never carry this likeness too far; it decreases their ferocious dedication. Understood in terms of Palestinian suicide bombing terrorism against Jews, this evokes a clear paradox. The Arab terrorists must acknowledge that their intended Jewish victims are also human, but just barely.

For the Palestinian terrorists, sacrificial violence against Israel has two categories of victims. One category is the "vile, infidel Jew." The other is "glorious martyr" who kills despised Jews and who, by it, earns eternal fame by "dying for the sake of Allah." This "martyr" need not fear personal death in sacrificing himself as a suicide. On the contrary, by choosing to die this way, he actually buys himself free from the penalty of dying: "Do not consider those who are slain in the cause of Allah, as dead," says the Koran. "They are living by their Lord."

"Strive for death, and you will receive life," believes the Palestinian terrorist who would sacrifice himself as well as the hated "Jew plunderer." A sequence of articles on Palestinian "martyrs" (the Shuhada) in Al-Istiqlal implicitly recognizes a basic human sameness but also believes a large difference from the Jews, whom they affirm to seek life at all costs. In essence, being unfounded and the most ironic separation, the main reason of the "suicidal" Arab and Islamic terrorist is to avoid death. This "martyr" focuses on his important issue and belief of earning a "seat in Paradise" and being saved "from the torture of the grave."

The most basic facts about our current clash of civilizations remain unrecognized. The true rationale of Arab or Islamic suicide bombing terrorism does not lie in politics. It is, more than anything, an example of religiously based blood sacrifice, a primal ritual designed to enlist divine support in Jihad. Expressed in the recurrent slaughter of carefully selected innocents, especially Jews, suicide bombing is never about land or rights or justice or national self- determination. For the pious murderer, it is a distinctly private and irremediably cowardly take on to secure both public flattery and personal immortality.

Terrorism has been with humanity since societies started having arguments. Infected corpses were used long before the germ theory of disease. Biological toxins extracted from plants were used to poison wells and assassinate leaders as far back as any history can trace.

Here are some examples of terrorism over the past three hundred years.

Eighteenth Century

- Infected corpses were used by the Russians against areas held by Sweden.

- The Puritans performed many acts of violence against other religious groups, especially the Catholics and Quakers, prior to and after the Revolutionary War.

- Organized violence against government taxation included Shay's Rebellion (1786) and the Whiskey Rebellion (1791).

- British officials provided blankets from smallpox patients to Native Americans.

Nineteenth Century

- There was the planned assassination of Tsar Alexander II.

- Twelve additional assassinations of public officeholders took place following that of President Lincoln.

- The Ku Klux Klan began acts of terrorism.

- In 1886, a peaceful labor rally at the Haymarket Square in Chicago was disrupted when an unknown person threw a bomb, killing seven.

- Catholic churches were burned in Boston, Philadelphia, and other cities from the 1830s to the 1850s.

- Industrial Workers of the World (IWW) activists use three thousand pounds of dynamite to blow up the Hill and Sullivan Company mine at Wadner, Idaho, along with a boardinghouse and bunkhouse.

Twentieth Century

- The 1950 assassination attempt on President Truman by Puerto Rican nationalists resulted in the death of one DC police officer during a gun battle outside the Blair House.

- In 1954, five members of Congress were wounded by gunfire during an attack by Puerto Rican nationalists;
- One of forty-nine bombings attributed to the Puerto Rican group FALN between 1974 and 1977 was the Frances Tavern bombing in New York City. Four people died in this event.
- In 1975, a bombing by Croatian nationalists killed eleven and injured seventy-five at LaGuardia Airport in New York City;
- In 1976, Orlando Letelier, a former Chilean ambassador to the United States, was killed, along with one of his associates, in a car bombing in Washington, DC.
- In 1981, a man was killed when a bomb planted by a group calling itself the Puerto Rican Armed Resistance detonated in a men's bathroom at Kennedy International Airport in New York City.
- During the latter half of the twentieth century, there were numerous airline hijackings and bombings.

As you can see, extremists have used property destruction and violence to produce fear and compel societal change throughout history. However, one major change has happened recently. Before modern times, terrorists usually granted certain categories of people (for example, women, children, clergy, elderly, infirm,) immunity from attack. Like other warriors, terrorists recognized innocents, people not involved in conflict. For example, in late nineteenth-century Russia, radicals planning to assassinate Tsar Alexander II, the leader of Russia, failed several planned attacks because they risked harming innocent people. Historically, terrorism was direct; it intended to produce a political effect though the injury or death of the victim.

Events such as the World Trade Center and Oklahoma City incidents and various clinics' bombings are designed to create a public atmosphere of anxiety and undermine confidence in government. Their unpredictability and clear randomness make it almost impossible for governments to protect all potential victims. Modern terrorism offers its practitioners many advantages. First, by not recognizing innocents, terrorists have an infinite number of targets.

They select their target and decide the best approach to attack. The range of choices gives terrorists a high likelihood of success with minimum risk. If the attack goes wrong or fails to produce the intended results, the terrorists can deny responsibility.

Along the planning logic of developing an antiterror strategy lurks the confrontation with the reality of terrorism as a social cancer. When women may be hiding explosive devices within their bodies, we must realize that we are no longer dealing with outlaws but with an army of great faith in the righteousness of their cause. Most outlaws understand they are on the wrong side of the law when arrested; they may even show remorse. However, Hezbollah, Hamas, PLO, Islamic Jihad, and Al-Qaeda believe firmly in the righteousness of their cause. This belief intensifies their danger. They serve Allah and will be rewarded in paradise.

We refuse to recognize what we already know about suicide bombers. They have a tendency to be Muslims between the ages of 18–45, and they have prayed in mosques, heard the sermons of imams, and watched their children play at Hamas kindergartens and summer camps. Further, when we have confronted terrorists, we continue to hold on to the belief that terrorists are merely criminals. Like the Schutzstaffel guards at Auschwitz are loving sons and fathers, never accused of any crime, they are often found as honest individuals. However, they do believe in an evil ideology.

Many apologists and propagandists argue in defense of Islam, that it was once, perhaps, a socially acceptable religion some centuries ago. These apologists employ all the buzzwords that would evoke the sympathy of the American Civil Liberties Union. They accuse us of being Islamophobes. So for the sake of historical accuracy, I agree that classical Islam had some socially valuable ideas, most of which, according to the eminent historian Will Durant, came from Judaism.

The "enemy of my enemy is my friend" goes to an old Arab expression. Even if these bombs go off in my neighborhood, the real target is not me; it is Israel, America, Christians, and Jews. Arabs accept these incidents with a sad composure; they are "friendly fire" tragedies. So Arabs in Gaza or Ramallah can be cheated and abused by Arafat, yet they remain focused on their real enemy, which is Israel. These terror incidents are no more than matters of internecine rivalries, no different in substance than Republican and Democrat or Labor and Likud debates. So we look in amazement at Saudis and Iraqis while they accept these bombings with little more than a shrug.

We are not chasing criminals but defending against enemies ready to destroy our families all in the name of perverse morality. Just look at their vocabulary. For example, Nakba refers to what they call the humiliation of 1948. Exactly what happened in 1948? The Arab world declared their purpose to drive the Jews into the sea.

Political correctness has become the central value of our culture. It proclaims the negative value of false equality. There is no longer right and wrong, good and bad. Suicide bombers are militants, and murderers are freedom fighters. A piece of the Arab world that lost its chance to live freely alongside a Jewish state back in 1948 has mutated into a moral cancer metastasizing within the body politic of the civilized world. We must share the medical metaphor with some Arab writers. It is not the "occupation" that causes terrorism; it is Islamic rhetoric and ideology, a worldview that is at first tyrannical.

For over two centuries, we have believed in and defended democracy. What if our enemies used democracy as a weapon against us to win the fight? At best, I think that Jefferson, Rousseau, and Locke imagined democracy as a way of ensuring liberty. Beyond their interest in this as a social or political organization, they shared a commitment to a higher order. Democracy is a classification through which we govern ourselves; liberty is a transcendent value.

Our fight against terror is attributed to our moral confusion between civil rights and security. Along this issue, you see philosophy and agenda of corporate America that further complicates matters. In a letter to Horatio Spafford, Thomas Jefferson wrote that merchants had no country—"The mere spot they stand on does not set up so strong an attachment as that from which they draw their gains." Alternatively, as a friend of mine put it, "just follow the money."

Bush spoke of a war against terror that may continue for decades. Eventually, enough Americans may realize the mind in control of these terrible acts is the Islamic mind. The British historian R. G. Collingwood once defined history as the life of mind. Essentially, history is not the succession of events. Events are merely the external expressions of the action of mind. We must realize the war will not be won by soldiers alone.

The continuing war between the Arabs and Jews points to an unavoidable Arab victory unless Jews and Christians attack the ideological foundations of terror. We cannot eliminate the terrorists as long as the mosques and Hamas kindergartens perform freely. America's war against terror, which we now are fighting, may already have been lost. If we are winning the war, realize it can only be won when you set free the minds and hearts of a billion Muslims whose culture and morality celebrate murderers as martyrs. Fighting alongside our infantry and airmen must be the teachers and spiritual leaders of America and Israel.

There are three perspectives of terrorism: the terrorists, the victims, and the general publics. The phrase "one man's terrorist is another man's freedom fighter" is a view terrorist themselves would accept. Terrorists do not see themselves as evil. They believe they are legitimate combatants, fighting for what they believe in by whatever means possible. A victim of a terrorist act sees the terrorist as a criminal with no regard for human life. The general public's view is the most unstable. The terrorists take great pains to foster a Robin Hood image in hope of swaying the general public's position toward their cause. This compassionate view of terrorism has become an important part of their psychological warfare and needs to be countered energetically.

Can an individual truthfully be your enemy? If you look at the war we're fighting, as well as other countries that have been fighting for centuries, you might think so. When you consider the number of people who get in your way for no other reason than to cause you grief, if they're not your enemy, they sure do a good imitation of one. Even Paul said in 2 Corinthians 12:7, "And lest I should be exalted above measure through the abundance of the revelations, there was given to me a thorn in the flesh, the messenger of Satan to buffet me, lest I should be exalted above measure."

The apostle was troubled by what felt like a thorn in his flesh. What he meant upon stating "*thorn in my flesh*" is one of many speculations among biblical scholars. Some say it stands for a great sickness, trouble, or temptation. Others say it represented false apostles who were manipulative and sought him for their own selfish agendas. I believe the latter more closely fits within the context of Paul's writings. People just got on his nerves because they were not honorable and godly, but arrogant and evil.

I'm sure you know a few folks who fit the description of arrogant and evil, and you may even consider them to be your enemy, but there is something that you should always bear in mind. When you declare within yourself that you will without question sell out for Jesus Christ, when you fully persuade yourself that what God said He will do for you, He unconditionally, beyond any shadow of doubt will do it, you need to realize the reality that anyone can potentially become your enemy.

> For we wrestle not against flesh and blood, but against principalities, against powers, against the rulers of the darkness of this world, against spiritual wickedness in high places.

Ephesians 6:12 is about readiness. It's telling us—rather forthrightly, I might add—that our real enemy isn't another human being but one of infinitely more cunning and craftiness. This verse suggests to us the person who we believe is our enemy is nowhere near as diabolical and deadly as the devil. You see a human being can only visually find out whether you are wearing protective gear or have a weapon of some kind, but the devil is not so much concerned with that. The method he uses most is that of attacking you from the inside out. The first attack he is going to do is to mess with your mind so he can get you to doubt God.

He cannot read your thoughts, but he can without doubt read your actions. How does he do that, you might ask? Well, that's a good question, but before I answer it, you need to understand something with crystal clarity. Many people say, "God used me to do this" or "God used me to do that." I know what they mean; they mean to say the power of God was at work within them. This is a far better description than to suggest God uses people because to say that He will use us suggests that God will take over our free will, and we know that He would never do that. God doesn't use anybody. Only the devil uses people. He takes over their minds because they allow him to, and then he uses them to become a stumbling block in the lives of born-again believers.

How does the devil read your actions? Well, one of the ways he does it is through your so-called enemy. He uses your enemy to gauge whether your spiritual armor is intact, loose, or missing. If he can get a reaction from you at what he throws your way, then he knows your weakness, and that weakness gives him just the entrance he needs to continue to run you in the ground. So listen, don't stand for this; stand for Christ!

Jesus Christ has taught us how to make our enemies our footstools. This doesn't mean that they are beneath us, or making them to be submissive or subservient toward us. That can positively happen; we need to seek a better cause. Those you call your enemies are slaves to a tormenting master who uses them and, after he finishes using them against you, will chew them up and spit them out. This difficult task. You need compassion through God for those who come against you. Your desire must be to see them saved and filled with the glorious light of Christ.

Understand that if anyone is getting the best of you, they are only pricking you in the side because some part of your armor is missing, exposing your sore spot. Go search for your missing piece and purpose and put it on with the love of Christ. Always remember, you are more than a conqueror; nothing will stop you. God has etched your victory in stone, so don't act as if he hasn't. Deal with the enemy by displaying the strength of God's love through the example of Christ; that's the only way to kick him to the curb.

We have come to recognize individual spirits by the repeating pictures or visions given by the Holy Spirit to pinpoint a certain entity. For instance, if the Spirit shows us a mushroom cloud, as in the wake of the atom bomb, we know we are dealing with a spirit of insanity. Our flesh feels irritable when this spirit is around us. We may experience symptoms of panic and unusual fear and other issues, or our thoughts may race. Seeing all sensations are fruits of the insanity spirit being in our atmosphere. Therefore, even if we are ministering to a person who seems individually in control, they war against our flesh and, with the vision the Spirit gave us, do not fool us.

When we know the lifestyle of someone well enough to know they have allowed cracks to form in them through the works of the flesh, encourage them to repent. Go through deliverance, cleaning up—perhaps even getting baptized again—defeating the enemy, and being released from their lives. Without change, there is no way they can be overcomers and enter the kingdom. When they are at war against Satan, perhaps they are so bound up that they can't even see the enemy working in them. They may defend the right to be who they think they are and fight to continue to do what they want to do. This causes grief to the godly around them, but hard as it can be to deal with sometimes, just knows it's still not them.

Do combat against the spirit controlling them. For instance, we can pray their home environment clean, isolating them in a pure atmosphere if possible. Sometimes when we block outside entities from interfering and lock up a person with their pet demons in a spiritual cell of purity and righteousness, as you might say, they get a better perspective of themselves. They see who they have become and want rid of what has caused their condition. This is especially the case when principalities and powers of darkness have had real strongholds over them. We can see these evil forces if we are detecting, wrecking their lives, health and chances of being happy.

Satan will do his best, however, to lead those newly revived from the watery grave of baptism into destruction so they cannot be a part of this end-time army. He will try to send them back to the spiritually dead church where they were no threat to him. However, if they are noticing ones and obedient to the voice of God for direction, they will themselves take advantage of God's deliverance ministry, wipe out the enemy from their lives, and encourage others to do the same so the enemy is not controlling them. The light of the obedient will shine clearly for others to follow as they remain faithfully in God's army to defeat the body of Satan or Antichrist.

They will forever be a part of the victorious church eternal and will walk into eternity as a member of the body of Christ, binding every evil spirit plaguing humanity. Time will come to an end, and graves will be empty. All creations that are now corruptible will become incorruptible, and mortal will become immortal, bringing to pass to fulfill the last statement of Jesus as a man on the cross, "It is finished!"

8

Life as a Prisoner of War

FOR MOST OF human history, depending on the culture of the victors, enemy combatants on the losing side in a battle who had surrendered and been taken as a prisoner of war could expect to be either slaughtered or enslaved.[1] The first Roman gladiators were prisoners of war and were named according to their ethnic roots such as Samnite, Thracian, and the Gaul (Gallus).[2] Homer's Iliad describes Greek and Trojan soldiers offering rewards of wealth to opposing forces who have defeated them on the battlefield in exchange for mercy, but their offers are not always accepted

Typically, little distinction was made between enemy combatants and enemy civilians, although women and children were more likely to be spared. Sometimes, the purpose of a battle, if not a war, was to capture women, a practice known as raptio; the rape of the Sabines was a large mass abduction by the founders of Rome. Typically, women had no rights and were held legally as chattel.

The earliest known purposely built prisoner-of-war camp was established at Norman Cross, England, in 1797 to house the increasing number of prisoners from the French Revolutionary Wars and the Napoleonic Wars. The average prison population was about 5,500 men. The lowest number recorded was 3,300 in October 1804, and 6,272 on April 10, 1810, was the highest number of prisoners recorded in any official document. Norman Cross was intended to be a model depot, providing the most humane treatment of prisoners of war. The British government went to great lengths to provide food of a quality at least equal to that available to locals. The senior officer from each quadrangle was permitted to inspect the food as it was delivered to the prison to ensure it was of sufficient quality. Despite the generous supply and quality of food, some prisoners died of starvation after gambling away their rations. Most of the men held in the prison were low-ranking soldiers and sailors, including midshipmen and junior officers, with a small number of privateers. About 100 senior officers and some civilians "of good social standing," mainly passengers on captured ships and the wives of some officers, were given parole outside the prison, mainly in Peterborough, although some further afield in Northampton,

Plymouth, Melrose, and Abergavenny. They were afforded the courtesy of their rank within English society.

At the start of the civil war, a system of paroles operated. Captives agreed not to fight until they were officially exchanged. Meanwhile, they were held in camps run by their own army where they were paid but not allowed to perform any military duties.[3] The system of exchanges collapsed in 1863 when the Confederacy refused to exchange black prisoners. In the late summer of 1864, a year after the Dix-Hill Cartel was suspended, Confederate officials approached Union general Benjamin Butler, Union Commissioner of Exchange, about resuming the cartel and including the black prisoners. Butler contacted Grant for guidance on the issue, and Grant responded to Butler on August 18, 1864, with his now famous statement. He rejected the offer, stating, in essence, that the Union could afford to leave their men in captivity, the Confederacy could not.[4] After that about 56,000 of the 409,000 POWs died in prisons during the American Civil War, accounting for nearly 10 percent of the conflict's fatalities.[5] Of the 45,000 Union prisoners of war confined in Camp Sumter, located near Andersonville, Georgia, 13,000 (28 percent) died.[6] At Camp Douglas in Chicago, Illinois, 10 percent of its Confederate prisoners died during one cold winter month, and Elmira Prison in New York state, with a death rate of 25 percent, very nearly equaled that of Andersonville.[7]

Under the Third Geneva Convention, prisoners of war (POW) must be:

- treated humanely with respect for their persons and their honor,
- able to inform their next of kin and the International Committee of the Red Cross of their capture,
- allowed to communicate regularly with relatives and receive packages,
- given adequate food, clothing, housing, and medical attention,
- paid for work done and not forced to do work that is dangerous, unhealthy, or degrading,
- released quickly after conflicts end; and
- not compelled to give any information except for name, age, rank, and service number.[8]

During World War I, about 8 million men surrendered and were held in POW camps until the war ended. All nations pledged to follow The Hague rules on fair treatment of prisoners of war, and, in general, the POWs had a much higher survival rate than their peers who were not captured.[9] Individual surrenders were uncommon; usually a large unit surrendered all its men. At Tannenberg, 92,000 Russians surrendered during the battle. When the besieged garrison of Kaunas surrendered in 1915, 20,000 Russians became prisoners. Over half the Russian losses were prisoners as a proportion of those captured, wounded or killed. About 3.3 million men became prisoners.[10]

Historian Niall Ferguson, in addition to figures from Keith Lowe, tabulated the total death rate for POWs in World War II as follows:[11]

	Percentage of POWs that Died
Soviet POWs held by Germans	57.5 percent
German POWs held by Yugoslavs	41.2 percent
German POWs held by Soviets	35.8 percent
American POWs held by Japanese	33.0 percent
German POWs held by Eastern Europeans	32.9 percent
British POWs held by Japanese	24.8 percent
German POWs held by Czechoslovaks	5.0 percent
British POWs held by Germans	3.5 percent
German POWs held by French	2.58 percent
German POWs held by Americans	0.15 percent
German POWs held by British	0.03 percent

Prisoners of war from China, the United States, Australia, Britain, Canada, India, the Netherlands, New Zealand, and the Philippines held by the Japanese armed forces were subject to murder, beatings, summary punishment, brutal treatment, forced labor, medical experimentation, starvation rations, poor medical treatment, and cannibalism.[12] The most notorious use of forced labor was in the construction of the Burma-Thailand Death Railway. After March 20, 1943, the Imperial Navy was under orders to execute all prisoners taken at sea.[13] According to the findings of the Tokyo Tribunal, the death rate of Western prisoners was 27.1 percent, seven times that of POWs under the Germans and Italians. The death rate of Chinese was much higher. Thus, while 37,583 prisoners from the United Kingdom, Commonwealth, and Dominions, 28,500 from the Netherlands, and 14,473 from the United States were released after the surrender of Japan, the number for the Chinese was only 56. Of the 27,465 United States Army and United States Army Air Forces POWs in the Pacific Theater, they had a 40.4 percent death rate. The War Ministry in Tokyo issued an order at the end of the war to kill all surviving POWs.

According to some sources, the Soviets captured 3.5 million Axis servicemen (excluding Japanese), of which more than a million died.[14] One specific example is that of the German POWs after the Battle of Stalingrad, where the Soviets captured 91,000 German troops in total (completely exhausted, starving, and sick), of whom only 5,000 survived the captivity. German soldiers were kept as forced labor for many years after the war. The last German POWs, like Erich Hartmann, the highest-scoring fighter ace in the history of aerial warfare who had been declared guilty of war crimes but without due process, were not released by the Soviets until 1955, three years after Stalin died.[15]

There were stories during the Cold War to the effect that 23,000 Americans who had been held in German POW camps were seized by the Soviets and never repatriated. This myth had been perpetuated after the release of people like John H. Noble. Careful scholarly studies have demonstrated this is a myth based on a misinterpretation of a telegram that was talking about Soviet prisoners held in Italy.[16]

Alistair Urquhart, a ninety-one-year-old veteran of the British Army's Gordon Highlanders, spent three and a half years as a prisoner of the Japanese—one of 80,000 Brits who surrendered after the fall of Singapore. "Just when you thought it couldn't get any worse," he recalls, "it did." Urquhart's gripping book, *The Forgotten Highlander*, recently released in the United States, recounts how he survived slave labor on the notorious Death Railway in Thailand; blindness and paralysis; and the sinking of his "hell ship" prisoner transport, followed by five days alone at sea. His skills, grit, and self-discipline, with bits of luck, ultimately brought him a full and satisfying life. But he was denied an army pension, still needs painkillers, and suffers lingering metabolic damage as a result of his ordeal.

> Two days before they came to the army headquarters at Fort Canning, the Japanese massacred 300 people at Alexandra Hospital—doctors, nurses, even patients on the operating table. When we were marched to Changi prison camp, they deliberately had us pass a gruesome sight and stench: Chinese heads mounted on poles and decaying bodies. Marching with us a distance away was a column of Chinese, who were later massacred. At Changi, we were hungry and sick, and men began dying, but the Japanese usually left us alone. But I was only there for three months; I was among the first draft of 600 men to go to Thailand to build the Death Railway and bridges over the Kwai.
>
> We had a half cup of boiled rice and a cup of water every day—for three and a half years. Can you imagine working in that tropical sun, hewing solid rock with pickaxes and shovels, on those rations? We wore Jap-happies—their kind of underwear—and nothing else. No hats. Bare feet. So we all soon were suffering from a host of diseases: malaria, dysentery, beriberi—that was a real killer. And depression.[17]

Those who haven't lived life as a POW can only imagine the brutality of that life through the stories they read. We hear all the time about the POWS in Vietnam, and some believe they still exist today. Even though we still see POWs today in the new war, how about looking deeper? We all have, at one point in our lives, been a POW is our spiritual walk.

The Bible reveals that the ministry of Jesus Christ is diametrically opposed to that of the devil and his agents. While the enemy and his agents ensnare and imprison people, including Christians, Jesus sets people free from all forms of captivity. This message is filled with the power to break satanic prison doors around your life and set you free so that you can enjoy your divine benefits in Christ fully. As you read it and pray the prayers that accompanied it with rocklike faith, God's power to set the captives free and deliver to the uttermost will come upon you. Remember that the Bible says that the Lord is that Spirit and where the Spirit of the Lord is, there is liberty.

> To open the blind eyes, to bring out the prisoners from the prison, and them that sit in darkness out of the prison house. (Isa. 42:7)
>
> When their eyes were opened by means of the Gospel sent among them, through the energy of the divine Spirit; for this is a work of almighty power and efficacious grace: to bring out the prisoners from the prison. They were concluded in sin, shut up in unbelief, and under the law, the captives of Satan, and held fast prisoners by him and their own lusts, under the dominion of which they were: and them that sit in darkness out of the prison house.

Satan wants to blind you to the truth and therefore defeat you overall. This evidently refers to a spiritual deliverance, though the language is derived from deliverance from a prison. It denotes that he would rescue those who were confined in mental darkness by sin and that their deliverance from the thralldom and darkness of sin would be as wonderful as if a prisoner should be delivered suddenly from a dark cell and be permitted to go forth and breathe the pure air of freedom. Such is the freedom which the Gospel imparts; nor can there be a more striking description of its happy effects on the minds and hearts of darkened and wretched people.

That thou mayest open the eyes of the blind. Here he explains more fully for what end Christ shall be sent by the Father, that we may see more clearly what advantage he yields us and how much we need his assistance. He reminds all men of their "blindness," that they may acknowledge it, if they wish to be illuminated by Christ. In short, under these metaphors, he declares what is the condition of men till Christ shines upon them as their Redeemer—that is, that they are most wretched, empty, and destitute of all blessings and surrounded and overwhelmed by innumerable distresses till they are delivered by Christ.

What a grievous affliction is blindness! It was no frivolous boon which Christ, in the days of His sojourn on earth, thought proper to confer, when, in the external sense, He opened blind eyes. In the paragraph of which the text is a part, Jehovah is describing the Messiah in His spiritual character and work, and great as the marvel of removing natural blindness was and great as similar miracles were which Christ performed, their principal value consisted in their being symbols and pledges of those spiritual operations which He could accomplish on the souls of men.

To bring out the prisoners from the prison. This means to free poor souls from the tyranny of sin and terror of hell. This should make us say to Christ, as one did once to Augustus for a deliverance nothing so great, "Effecisti, Ceesar, ut viverem et morerer ingratus." Let me do mine utmost, I must live and die in thy debt.

The prisoners, sinners, are taken captive by the devil at his will as we read in 2 Timothy 2:26, "And that they may recover themselves out of the snare of the devil, who are taken captive by him at his will). John 8:32, 36—compare this portion of Scripture with Isaiah 61:1, and both with Luke 4:17–21, where it is said to be fulfilled in and by Christ.

Okay, here is the kicker in this whole scene. Satan's power has standing in the spiritual realm, where he has access to the presence of God. The book of Job provides insight into the association between God and Satan. In Job 1:6–12, Satan stands before God and reports that he has been "walking up and down" on the earth (verse 7). God asks Satan if he has considered godly Job, and Satan immediately accuses Job of insincerity he only loves God for the blessings God gives. "Stretch out your hand," Satan says, "and strike everything he has, and he will surely curse you to your face" (verse 11). God grants Satan permission to affect Job's possessions and family, but not his person, and Satan leaves.

So when we read this, it tells us one thing, we are only prisoners of Satan because we allow it, not Satan. Does God have that much faith in you as He did Job? Most people will not think outside the box, but not only did Job have faith in God, but God Himself had faith in Job. See, we at times get our feathers ruffled because we go through challenges in our lives. We need to stop and realize that God has faith in us to do the right thing. What an awesome feeling that God the Father has faith in us!

The bottom line is when we become POWs in life fighting in the wars, serving our country, it is not because of our wrongdoing. We are captured of bad orders, mechanical issues, or just being in the wrong place at the wrong time. We, in a way, cannot help this; however, with our Christian life, it is up to us. We have full control of what happens, and it is up to use to choose very carefully.

9

The Final Battle

THE APOCALYPSE, JUDGMENT Day, the End of Times, and Armageddon, through the centuries, have been absorbing the human race with making out Armageddon is a spirited role. The *battle of Armageddon* or just *Armageddon* is a common doomsday term. While people claim to use it based on Bible facts, the use of *Armageddon* unusually often ignores what the Bible itself plainly says about it, replacing the true meaning of Armageddon by the secular idea of humanity destroying itself. The word is now one and the same with nuclear war, natural disasters, impact from a comet or an asteroid, and any number of creatively conjured existential threats. Nowhere does the Bible identify a single one of these events with Armageddon.

Armageddon—the name of the place that will be the scene of the final battle, predestined by God, between good and evil in the earth at Christ's Second Coming, this final battle being greatest of all battles in history fought in this world. Christ's standing on one side, and the Antichrist standing on the other aided by the devil, Satan, his false prophet, and the armies of the confederation of nations that will align them with Antichrist at Armageddon. The dragon is the devil, casting him down to earth from heaven. With Antichrist and the false prophet, the devil will make war on Israel in the second three and a half years of the tribulation. The woman in Revelation 12 symbolizes Israel. The third part of the stars in heaven in verse 4 refers to the number of angels also cast out of heaven down to earth with the devil.

To get the timing for the battle of Armageddon, we need to realize the book of Revelation is revealing us the sequence of events. The book is in general opening a vision of a scroll sealed with seven seals, which, sequentially, after opening the seventh and final seal, John sees seven angels with seven trumpets. After the seventh angel blows his trumpet, John sees seven angels with seven bowls of wrath. It is at the end of the sixth bowl of wrath the events leading to the battle of Armageddon begin to happen.

They feed the deceived humans a constant lie through the media to make them assume the attacking forces are aliens coming from outer space. Rightly so, but these "aliens" are Christ and His armies. The humans on earth fighting God Himself, just as they did during the first coming of Christ, in which they crucified Him. The difference now is that Christ is not coming to earth as the Lamb to prepare Himself to be slaughtered; He is returning as King of kings and Lord of lords, and He will win this battle against Satan.

Some Christians believe inaccurate beliefs about Armageddon. Many correctly view Armageddon as the gathering place of the Antichrist and his armies before the Glorious Appearing of Jesus Christ, but they mistake to intend this gathering. Commonly held beliefs are these armies gather at Armageddon to battle with one another or wipe out the Jewish people. Only through the physical intervention of Jesus Christ is humanity saved from either destroying all life or the Jewish person. Neither motive is correct. The Bible clearly states why these armies gather at Armageddon, and once we understand why, it is easy to see why most Christians find it difficult to imagine.

Armageddon—a real place, one that seen more violence and bloodshed than any place on earth. During the past four thousand years, there were at least thirty-four bloody conflicts have already been fought at the ancient site of Megiddo and bordering areas of the Jezreel Valley. Egyptians, Canaanites, Israelites, Midianites, Amalekites, Philistines, Hasmonaeans, Greeks, Romans, Muslims, Crusaders, Mamluks, Mongols, French, Ottomans, British, Australians, Germans, Arabs, and Israelis have all fought and died.

The Bible teaches that before the battle of Armageddon, the times will be so terrible on earth that humanity will turn to anyone to rescue them except God. At first, the united world armies are reluctant to take on the forces of God led by the Lamb of God, which is Jesus.

Revelation 6:14–16 says,

> And the heaven departed as a scroll when it is rolled together; and every mountain and island were moved out of their places. And the kings of the earth, and the great men, and the rich men, and the chief captains, and the mighty men, and every bondman, and every free man, hid themselves in the dens and in the rocks of the mountains; and said to the mountains and rocks, Fall on us, and hide us from the face of him that sitteth on the throne, and from the wrath of the Lamb.

Before destroying Babylon, Satan sends out "spirits of demons, performing signs, which go out to the kings of the whole world. They gather them for the war of the great day of God, the Almighty…And they gathered them together to the place which calling it in Hebrew is Har-Magedon." The miracles of Antichrist and his henchmen are enough to convince the world leaders that Antichrist has the wits and power to overcome God.

They are under incredible pressure. First, the economic, military, social, physical, and political world is in chaos.

Second, they refuse to turn to God. Third, these charismatic, intellectual, crafty, and influential figure cons them into believing he can defeat God with their help. Times are terrible, and they must act quickly. However, he is leading them into a horrifying ambush, the great trap called the battle of Armageddon.

The blood of the armies of Antichrist, who Christ will destroy at Armageddon, will flow for two hundred miles, nearly three hundred and twenty kilometers at a depth of roughly 1.2 meters to the horses' bridles. Armageddon is the great winepress of the wrath of God in Revelation 14:19. It will take seven years to burn the weapons of war used by Antichrist's armies as firewood. It will take all of Israel seven months to bury the bodies of those killed in the battle, despite the fowls of the air and the beasts of the field will also eat the dead.

Destroying much of Palestine, as too will much of the world in the battle of Armageddon and the events preceding it, immediately before Christ's Second Coming. This earthquake will be the most powerful and devastating earthquake in the history of the world. It will split Jerusalem into three parts, bring cities around the world crashing down, and outright destroy the city of Babylon. Radically alter the earth's topography, moving every mountain and island from their position. The mountains will figuratively disappear. This will happen contemporaneous with Antichrist's armies invading Israel immediately before Christ's Second Coming and the battle of Armageddon.

The Old Testament lists the wars and conquests of ancient Israel that God used to drive out the adulterous nations in Canaan to fulfill the promise made to Abraham:

> And I will give unto thee, and to thy seed after thee, the land wherein thou art a stranger, all the land of Canaan, for an everlasting possession; and I will be their God. (Gen. 17:8)

With the help of God, Israel fought some incredible wars. Remember, Joshua prayed during the battle, for the sun, and the moon to stand still, until Israel thoroughly defeated the five Amorite kings (see Josh. 10:13–14). Gideon's famous army of three hundred defeated the Midianites. Armed with a ram's horn and a clay jar, with a smoldering torch inside (see Judg. 7:16–21), Jehoshaphat went to battle with a choir, singing, "Praise the Lord; for his mercy endureth forever" and defeated the armies of Edom, Ammon, and Moab (2 Chron. 20:21).

Before these wars, a war in heaven took place between Christ and Satan.

> And there was war in heaven: Michael and his angels fought against the dragon; and the dragon fought and his angels. (Rev. 12:7)

In the last century alone, the United States involved itself in major wars, two world wars, developed the atomic bomb, engaged in the arm's race against the Soviets, fought

Seeing War Through God's Eyes   171

in the Korean and Vietnam wars, fought in the Gulf War, and is now at war on terrorism in Iraq and Afghanistan.

These facts cause us to wonder if a global war takes place wiping out the human race. Christ said while on earth, "And ye shall hear of wars and rumours of wars: see that ye be not troubled: for all these things must come to pass, but the end is not yet" (Matt. 24:6). Could Christ have been hinting about Armageddon mentioned in Revelation 16:16, "And he gathered them together into a place called in the Hebrew tongue Armageddon"?

The battle of Armageddon will need to begin near the end of the seven-year tribulation period. A sin overpowered "exponentially catastrophic" world's remainder (driven by demonic forces) will march to war against Almighty God and His eternal control over His creation (Joel 3:2, 12).

What happens during the battle of Armageddon and the location the battle takes place? Armageddon will be a real war; there is no disputing that. If we do a search on it, one will find most websites declaring that it will be a physical war between the nation of Israel and other nations. Some think maybe Russia, Syria, or Iran, and yet other websites will describe a fictitious event such as an asteroid hitting earth or something to that extent. As we learn from Israel in the end times, most Christians are focusing their attention on the literal nation of Israel.

The physical nations of Israel are no longer God's chosen because God no longer regards the flesh, but anyone who comes to Christ, whether Jew or Gentile, about them as spiritual Israel, and God's chosen. So when the book of Revelation talks about Israel, it is talking about spiritual Israel, not the physical nation. We can see the promises found in the Old Testament to the nation of Israel were conditional.

This overall end of rebellious humanity of the ravished world now will make their final delusional destination, clawing and crawling to war against their Creator. So focused, driven, closed with such exponential distinguishing, vindictive hate, that seven years of the horrific significances of sinful man with the divine judgments of Almighty God will not detour these guy's hate. They are overshadowing need to make their slow, and growing fast wrong utopia, and seek the highest purge Almighty God from His creation (Rev. 9:20–21, 16:9, 11).

No wonder Almighty God will laugh His head off at the full-blown insanity of the nations of this period (Ps. 2). Just before this battle's beginning, as the world's worldwide armies, under Lucifer's demonic leadership and the anti-Christ's despotic mandate, which comprises the ten regions of the world, they will advance their way through seven years of international catastrophic obstacles and tribulation to war with Almighty God.

These people from the ten regions of the Antichrist will advance through the seas and rivers of blood which by Almighty God was cursing (Rev. 16:3–6, 17:13–14). They will advance through the catastrophic global landscapes covered with the horrific carnage of the global war. They will advance through the catastrophic obstacles of the natural and supernatural disasters inflicted by Almighty God in His wrath. In their folly and addiction, they will never stop marching until they have arrived at their destination and show down. They will not let go of their hatred and rage as it thoroughly consumes them.

They are all fully unrepentant and have a "Pharaoh's heart of rebellion and stone." As this global horde continues their raging advance to Megiddo among all Almighty God's divine obstacles, Almighty God will unleash His final twenty-first divine judgment on this raging demonically induced advancing global host. In this final "bowl judgment," Almighty God will proclaim to all His creations, "It is done." He will cause every global tectonic fault line to move out of their place and institute a global earthquake that is incomprehensible.

This earthquake will be so exponentially disastrous, so exponentially horrific, so exponentially ravishing that it will transform and ravish the entire world's continents, causing tsunamis, wiping out all islands, coastlines and mountains. It will destroy all the world's magnificent cities, technologies, and wipe millions of human lives out (Isa. 24:5–6).

This master of intrigue, the Antichrist, takes on the Prince of princes in battle. Like the Tower of Babel generation, his arrogance will extend all the way into the heavens, where he will take to take over the glory and authority of God Himself. Note reading the last verse of this passage, you see the power of the Antichrist will not be able to break. No one will be able to make war against him. At least no human will be able to make war against him. When the Antichrist engages the Lord Jesus Christ in battle, he will be utterly destroyed. I believe that Jesus will destroy the armies of Satan with the two-edged sword described earlier.

The result of Armageddon is described in the book of Revelation, stating their destruction into the hands of Jesus Christ—lying dead bodies in the valley as a feast for the birds of the world, then capturing the beast and false prophet, throwing the both into the lake of fire, where they will experience everlasting torment. As we see Satan as a being of darkness, we realize he is also a being of light. He started the first war in heaven when he opposed God's philosophy of divine order. The divine war between the two forces of light and corrupted light is taking place on all levels of reality. Know that the actions between God and Satan in the flow of events from beginning to end of time. We have to discover strategies found in the art of war. The genius contained within solves many of the mysteries present.

The art of war is the art of deception, and the art of winning bases all warfare on deception. In war, practice dissimulation, and you will succeed. Dissimulation is a form of deception in which one disguises the truth. Dissimulation commonly takes the form of disguising one's ability to gain the element of surprise over an opponent. The Bible says that Satan deceives the whole world for fulfilling his agenda. He hides portions of the truth and mixes in error to create lies and illusions for others.

God, being all wise, also uses dissimulation on Satan. He revealed that Christ would be the Messiah of the world but did not reveal exactly how it would happen. Satan had the intent to destroy God's plan by killing Christ. He thought that the Son of God had come to save the human race and the earth by ruling as their king. Christ came to save the human race by dying for the sins of the world. This is the hidden wisdom of God that the devil prince knew, or else they would not have crucified the Messiah.

I'm grateful that we have a loving Heavenly Father, who, through His Son, has opened the future to us so we do not have to live one's lives with confusion, keeping everything clear about the final war and the results. Men will continue to mediate and break peace until Christ's returns. There will be no lasting peace in the Middle East or globally until the Prince of Peace returns to the Middle East in the clouds of glory.

> And he shall judge among the nations, and shall rebuke many people: and they shall beat their swords into plowshares, and their spears into pruninghooks: nation shall not lift up sword against nation, neither shall they learn war any more. (Isa. 2:4)

Something that many Christians do is that they mistakenly take certain words in Bible prophecy and apply them accurately, like the mark of the beast. Many think that this is going to be a physical, literal mark on our hand or forehead. They say the word means a mark, so it must be a physical mark. And yet the word beast in Revelation 13 means a beast. So are we now to believe that we will have literal beasts causing the world to worship them? No! We are to apply them in a spiritual sense, not physical, because there are many examples in Revelation that are symbolic, not literal. So if we look at the root meaning of *Armageddon*, we will see what this, in reality, means. The Hebrew term *har-megiddon* means "a mountain of crowds or congregation." Another mount in the Bible is mentioned, Mount Zion, and the place where God dwells:

> Beautiful for situation, the joy of the whole earth, is mount Zion, on the sides of the north, the city of the great King. (Ps. 48:2)

After the battle of Armageddon at the beginning of His millennial reign, Christ will judge the nations. (Matt. 25:31–46) Calling the judgment of the Nations, although, in fact, it will be individual people who will be judged, Jesus separates them from one another, into sheep and goats. (Matt. 25:32–33) The sheep represents those individuals who will go into the eternal Kingdom. (Joel 3:2, 12) The goats represent those who will be cast down to hell. (Matt. 25:46) All nations that will be gathered before Jesus in Matthew 25:32 are Gentiles, who survive the tribulation. Nations in Greek also mean Gentiles. Their judgment takes place before Christ setting up. His millennial kingdom on earth, to find out who of them will go into the kingdom.

The basis of the Gentiles' judgment will be their failure to extend mercy to the Jewish believers during the tribulation (Matt. 25:34). This can never be construed as teaching salvation by works, that these Gentiles were saved because of their kindness to the Jews. That would contradict the testimony of Scripture. It simply means that those Gentiles' kindness to the Jews during the tribulation reflected their love for Christ. That is what saved them. Furthermore, although the application of this teaching is toward the Gentiles, who survive the tribulation, the teaching is relevant to Christians in all ages. Believers today must also extend mercy to the least of God's children, whether they are Jews or Gentiles.

10

Destination Paradise

CHRISTIANS OFTEN SAY that paradise will be a spiritual place, and they often argue against Islam because the Islamic heaven includes physical qualities. However, when one reads the Gospel, the Gospel of Matthew, specifically, one will see that Jesus taught that heaven is physical, or partially physical. As he says,

> Ye have heard that it was said by them of old time, Thou shalt not commit adultery: But I say unto you, That whosoever looketh on a woman to lust after her hath committed adultery with her already in his heart. And if thy right eye offend thee, pluck it out, and cast it from thee: for it is profitable for thee that one of thy members should perish, and not that thy

> whole body should be cast into hell. And if thy right hand offend thee, cut it off, and cast it from thee: for it is profitable for thee that one of thy members should perish, and not that thy whole body should be cast into hell. (Matt. 5:27–30)

Well, first off, heaven is a place for the saved *only*. There will be no God rejecters in heaven. If you wish to go to heaven, then just accept the Lord Jesus Christ as your personal Savior and Lord. That is the only way to go to heaven. What does the Bible say about heaven? Well, the Bible gives us some clues but does not go into great detail about the future destination of the saved. Why is that so? Well, I believe that heaven will be so great and glorious that our minds cannot contain the blessings and riches of heaven. It hasn't even entered into our minds, the good things God has prepared for His people in heaven.

So as you can see, Jesus says if your right eye looks lustfully at a lady, then pluck it out, because it is best if you just pluck your eye out and throw it away for the sin it committed rather than keep it and be thrown entirely into hell because of it! The same thing with your hand. If you do something bad with your hand, cut it off, because it is best to get rid of the bad sinful body part that committed a sin rather than having your whole body thrown into hell on account of one body part!

It's good to think about what paradise will be like. It's something to long for. There will be no more sickness, no more working for other people or doing their work. You will be doing useful work for yourself, able to choose what you do and not having governments controlling you and making you their slaves. Instead, we will be at the mercy of God. What does the Bible say that heaven looks like? Are there sufficient scriptures to tell us? Are the images of the saints playing harps while lying on the clouds of heaven accurate? We know for sure what heaven will be like?

The tiny chubby cherub angels playing their harps in heaven are not what heaven will be like or is like at this present time. Heaven exists today, but humans who are living on earth cannot yet see it, and the pictures that we have seen of it cannot describe it accurately in all its splendor, glory, beauty, and majesty. Those who have died in the faith in Jesus Christ are said to be present with the Lord, and since the Lord is seated at the right hand of the Father on His throne, they are in heaven right now. What this means is that heaven is for real; it does exist, and there are Christians who have died and gone to heaven and are now with Jesus Christ.

There are several people in the Bible who have been eyewitnesses to the heaven before they died. Elijah was caught up into heaven, taken up in a whirlwind, never tasting death (2 Kings 2:1–12). Like Elijah, Enoch, who was said to have walked with God, was taken up into heaven without ever tasting death. The apostle John saw heaven through an open door so that he might see and record in the book of Revelation some of the things he saw in heaven (Rev. 4:1).

I am sure you have read or heard about many who have died and seen heaven for themselves. They describe such a realm that is beyond description. Some who died and came back to life revealed seeing their lost loved ones in heaven, their family, friends, and some even their pets. This is difficult to prove, yet the sheer number of these who have seen heaven and came back from death is hard to refute. Only they themselves know this for certain. However, the Bible does describe some of what heaven is like in the Scriptures. It is, from the human standpoint, indescribable (1 Cor. 2:9).

In the last book of the Bible, the book of Revelation, we find God's final word on human history. And when we reach Revelation 21, God unfolds for us the establishment of His eternal kingdom heaven, paradise, eternity with Him. Up until the very end of the Bible, we know almost nothing about what heaven will be like. In fact, the last three chapters of Revelation tell us more about heaven than all the rest of Scripture put together.

In these final chapters, we are told there will be three heavens. The first heaven will include the atmosphere, including the clouds and oxygen we breathe. The second heaven is outer space, where all the planets, constellations, and galaxies exist. And beyond that, there is a third heaven. Paul gives us a slight hint at how incredible this third heaven will be when he says in 1 Corinthians 2:9, "But as it is written, eye hath not seen, nor ear heard, neither have entered into the heart of man, the things which God hath prepared for them that love him."

In Revelation 21, John goes on to tell us that when we get to this new heaven, we will be living for eternity in a new creation, a new city, a new community with a new constitution. The reference to the holy city as a bride, dressed up for a wedding celebration, symbolizes for us the intimacy we will have with God in heaven. And what this tells us is that heaven is, first of all, a relationship. And by using this illustration of a man and woman becoming husband and wife, God wants us to understand what our relationship with Him will be like in heaven. The most intimate relationship humans can share will only be a shadow of the intimacy we will have with God for eternity!

All humanity began in a garden but will end in this new city, prepared like no other city that has ever been built, a city prepared for you and me who have put our trust in Jesus Christ. What an amazing place heaven will be prepared by our Creator for His creation to live for all time! Heaven is something that is beyond our wildest imagination, something I pray you will look forward to expectantly every day!

There are mansions throughout heaven. According to Roy Reinhold, "These beautiful buildings are open to all, meaning that there are no front doors, so that anyone can come and visit." There is no crime; whatever belongs to you can't be taken. There are streets of gold, a river of the water of life as clear as crystal, and walls made of precious stones (Rev. 21:18–21).

The place where all of God's children are going to live with the Lord forever is not some fancy dreamland way off in outer space somewhere, but an even more amazing dream city that's going to come down from God, out of space, to a beautiful new earth! And God is going to come down and live with us and we with Him in that heavenly city (Rev. 21:1–3). *Webster's Dictionary* says that paradise is the Garden of Eden, in which Adam and Eve was placed immediately after their creation. Another of the *Webster's* definitions calls it a place of bliss, a region of supreme felicity or delight. Apparently, it is a Persian word having to do with delightful gardens.

When the United States won its freedom from the British Empire, I am sure that at first our country seemed like paradise. There were a lot of freedoms, the tax burden was greatly reduced, and it seemed as though hard work and moving west provided great economic opportunities for many people. However, the United States was never paradise. We had slavery; there was ethnic pride and prejudice, capitalism with its excesses made virtual slaves out of many employees. If the legends of the old west are at all accurate, many people did that which was right in their own eyes, and, to an extent, anarchy prevailed in many places. Most Native Americans who had lived here before the country was populated by Europeans certainly didn't understand the new European culture, and they lost their way of life. They obviously would not have considered this new republic to be paradise.

Now heaven is generally thought of as paradise. The thief on the cross had an eleventh-hour conversion and was promised paradise with the Lord. I don't know if we need to eat there, but since the Lord ate after the resurrection (Luke 24:43), I suspect that we can eat there, and the food will be delicious. There will be no sin, no disease, no death, no lying or cheating or stealing. Best of all, it will be a place of love because the Lord will be there. Heaven will be a place of good company, good food, and good things to see. The government will not be by the people, of the people and for the people, but instead, we will have a benevolent dictator ruling paradise. His name is Jesus.

The United States is the best place in the present world to live, but it is no paradise. Representative government is abused by those with money. The law isn't always applied fairly to all. Taxes have become burdensome. Our land is filled with violence and with drugs and with a lack of reverence for God. However, there will be a future paradise. The paradise lost because of sin is going to be regained because of Calvary. Not everyone will be there, however. Only those who would enjoy the company of the Lord and of Christians and of a sinless society will be there.

Truthfully, many people do not want to be in a place like that. It will take a new nature and a new birth to enjoy this new paradise. I have to admit that I don't want to go there today, but I am looking forward to living in this paradise. The real attraction is going to be the benevolent dictator who loved us enough to die for us at Calvary.

Heaven is real! Millions and billions of God's children who have already died and gone on to be with the Lord are already enjoying this wonderful, gigantic, golden city of love, which is on its way to earth now! It is a real place, our final home, where everyone who loves the Lord is going to enjoy forever! But I don't think you could really understand just how wonderful heaven will be unless you first know how wonderful you will be when you get there! When Jesus comes back soon, all of God's saved children will receive new superbodies that will fly up from off the earth to meet Him in the clouds! (1 Thess. 4:16–17)

We'll be like Jesus was when he rose from the dead! He said we are going to have bodies like His! (Phil. 3:21; 1 John 3:2) After He came back from the grave, He could appear or disappear, walk right through walls or locked doors, and fly from one place to another with the speed of thought! And with even more power than the angels of God! (John 20:26). And just imagine how wonderful our homes will be there inside that gigantic golden heavenly city! Jesus said, "In My Father's House are many mansions…And if I go and prepare a place for you, I will come again, and receive you unto myself; that where I am, there ye may be also" (John 14:2–3).

In heaven, there will be no power shortage, for God is all powerful; no bumping into things in the dark; no sleepless nights. There is no night there (Rev. 21:25). Nor will there be bills, hospitals, bank accounts. In fact, heaven will have no complaints, no back talk, no hate, no fear, no crime. Heaven will be a place to worship God. When Isaiah speaks for the Lord of the new heavens and earth, he prophesies that "all flesh shall come to worship before me" (Isa. 66:22–23). The apostle John also predicts the same thing, saying, "For Thou only are holy: for all nations shall come and worship before Thee" (Rev. 15:4, 21:24).

Surely, heaven will be a place of joy and productive work. We can be sure that in heaven we will be like God and Jesus, and they have always worked. Jesus said, "My Father worketh hitherto, and I work" (John 5:17). We find this same description of the new heavens and new earth in Revelation 21:1 and Isaiah 65:17. Isaiah describes it as a place of joy, where there will be no tears nor the sound of crying. This is in Isaiah 65:19 and also in Revelation 21:4 and 7:17. Isaiah continues to show us it will be a place where one can labor and see his labor bear fruit. This is in Isaiah 65:21–23.

Heaven will have a beauty beyond our imagination. The best John could do to describe heaven was to say it was like an entire city of transparent gold. He says, "The walls of the new city are precious jewels. The gates are twelve single pearls. The city itself and its streets are pure gold, yet like transparent glass." This is found in Revelation 21:18, 19, and 21.

And the whole city is lit by the glory of God. Heaven is a place of matchless blessings. Jesus has prepared a mansion for us there. Great treasure is stored up for us; in fact, the saints "will inherit all things" (Rev. 21:7). The leaves of the tree of life will guarantee our health. And the crystal pure water of the river of life will flow from the throne of God to the tree of eternal life. For the first time since the Garden of Eden, God's children may eat out of this tree in His paradise (Gen. 3:22; Rev. 2:7).

In heaven, we will be given new names, new imperishable bodies. And we will be clothed in the white robes of righteousness. We will be glorious living stones, pillars in the eternal temple of God (1 Cor. 15:52; 1 Pet. 2:5; Rev. 3:12).

Best of all, heaven with God and Jesus will be a place of love. God, in tender love, will wipe away all our tears (Rev. 21:4). The angels will rejoice as heaven fills with righteous Christians (2 Pet. 3:13). Those spirits that have hungered and thirsted to humbly live with God, those with pure hearts, those who have willingly endured persecution, and those who have lived for God's peace will live together in life and love forever and ever. This place is heaven. Will you be there?

Notes

Notes to Chapter One

1. Matthew O. Jackson and Massimo Morelli, rev. ed., "The Reasons for Wars: An Updated Survey (December 20).

2. Ibid.

3. Cram101 Textbook Reviews, *e-Study Guide for: Modern East Asia: Brief History by Schirokauer*.

4. Emmanuel Badawi, "Transcript of Vietnam War" (13 June 2013).

5. John Whiteclay Chambers II, ed. *The Oxford Companion to American Military History* (New York: Oxford UP, 1999).

6. Ibid.

7. President Bush's address to a joint session of Congress on Thursday night, September 20, 2001.

8. Yaron Brook and Alex Epstein, "'Just War Theory" vs. American Self-Defense."

9. Barbara Ehrenreich, *Blood Rites: Origins and History of the Passions of War*, 45.

10. Hussein Ibish, "Orwell Would Revel in 'Collateral Damage,'" *Los Angeles Times*.

Notes to Chapter Two

1. "'Just War Theory' vs. American Self-Defense," *The Objective Standard*.

2. Yaron Brook and Alex Epstein, "'Just War Theory" vs. American Self-Defense."

3. Ibid.

4. Ibid.

5. Ibid.

Notes to Chapter Three

1. Gleason L. Archer Jr., *New International Encyclopedia of Bible Difficulties*.

2. Max Lucado, "Why Does God Allow War," https://maxlucado.com/why-does-god-allow-war/.

3. Jerry Falwell, "God is Pro-War," http://www.wnd.com/2004/01/23022/.

Notes to Chapter Four

1. "Why Does God Allow War," The Family International, http://www.thefamilyinternational.org/en/viewpoints/views-issues/20/.
2. Ibid.
3. "Why does God allow the innocent to suffer?" http://www.gotquestions.org/innocent-suffer.html.
4. Martyn Lloyd-Jones, *Why Does God Allow War?*.

Notes to Chapter Five

1. Army News Service (Sept. 7, 2006), GlobalSecurity.org.
2. www.army-technology.com is a product of Kab.

Notes to Chapter Seven

1. Yaron Brook and Alex Epstein, "'Just War Theory' vs. American Self-Defense."

Notes to Chapter Eight

1. Jason Wickham, "The Enslavement of War Captives by the Romans up to 146 BC" (doctorate thesis, University of Liverpool, 2014).

2. Encyclopedia Romana copyright 1997-2015 James Grout

3. Roger Pickenpaugh, *Captives in Blue: The Civil War Prisons of the Confederacy* (2013).

4. "Myth: General Ulysses S. Grant stopped the prisoner exchange, and is thus responsible for all of the suffering in Civil War prisons on both sides," Andersonville National Historic Site, https://www.nps.gov/ande/learn/historyculture/grant-and-the-prisoner-exchange.htm.

5. Richard Wightman Fox, *National Life After Death* (January 7, 2008).

6. "Andersonville: Prisoner of War Camp-Reading 1," Nps.gov, retrieved November 28, 2008.

7. *Wikipedia*.

8. "Geneva Convention," Peace Pledge Union, retrieved April 6, 2014.

9. Geo G. Phillimore and Hugh H. L. Bellot, "Treatment of Prisoners of War," *Transactions of the Grotius Society* 5 (1919): 47–64.

10. Niall Ferguson, *The Pity of War* (1999), 368–69.

11. Niall Ferguson, "Prisoner Taking and Prisoner Killing in the Age of Total War: Towards a Political Economy of Military Defeat," *War in History* 11, no.

2 (2004):186; Keith Lowe, *Savage Continent: Europe in the aftermath of World War II* (2012), 122.

12. Terry McCarthy, "Japanese troops ate flesh of enemies and civilians," London: *The Independent* (August 12, 1992).

13. Nigel Blundell, "Alive and safe, the brutal Japanese soldiers who butchered 20,000 Allied seamen in cold blood" (November 3, 2007).

14. Simon Rees, "German POWs and the Art of Survival."

15. "German POWs in Allied Hands—World War II," Worldwar2database.com., July 27, 2011.

16. Paul M. Cole (1994) POW/MIA Issues: Volume 2, World War II and the Early Cold War National Defense Research Institute.

17. 2015 History Net, Where History Comes Alive—World & US History Online.

www.ingramcontent.com/pod-product-compliance
Lightning Source LLC
Chambersburg PA
CBHW071920290426
44110CB00013B/1422